Importance of volume-regulated anion channel subunit LRRC8A for hypotonic stress response and differentiation of human keratinocytes

vom Fachbereich Biologie
der Technischen Universität Darmstadt
zur Erlangung des akademischen Grades
Doctor rerum naturalium
genehmigte

Dissertation

von

Janina Trothe

Darmstadt 2020

zugl.: Darmstadt, Technische Universität Darmstadt, Dissertation

Bibliographic information published by the Deutsche Nationalbibliothek

The Deutsche Nationalbibliothek lists this publication in the Deutsche
Nationalbibliografie; detailed bibliographic data are available
on the Internet at http://dnb.d-nb.de .

ISBN 978-3-8325-5165-0

Logos Verlag Berlin GmbH
Georg-Knorr-Str. 4, Gebäude 10
D-12681 Berlin
Tel.: +49 (0)30 42 85 10 90
Fax: +49 (0)30 42 85 10 92
https://www.logos-verlag.de

Für Mama und Papa

„You can't teach experience." – Ray Hunt

Die vorliegende Arbeit wurde in der Arbeitsgruppe von Dr. T. Ertongur-Fauth der BRAIN AG Zwingenberg konzipiert und angefertigt und von Prof. Dr. H. U. Göringer am Institut für Molekulare Genetik der Technischen Universität Darmstadt betreut.

Referent: Prof. Dr. H. Ulrich Göringer
Korreferent: Prof. Dr. Gerhard Thiel

Tag der Einreichung: 05.11.2019
Tag der Disputation: 17.01.2020

Teile der Arbeit sind in folgende Veröffentlichung eingegangen

Trothe, J., Ritzmann, D., Lang, V., Scholz, P., Pul, U., Kaufmann, R., Buerger, C., and Ertongur-Fauth, T., *Hypotonic stress response of human keratinocytes involves LRRC8A as component of volume-regulated anion channels.* Exp Dermatol, 2018. **27**(12): p. 1352-1360.

1 Zusammenfassung

Die humane Epidermis ist die äußerste Hautschicht und erneuert sich durch Proliferation und Differenzierung von Keratinozyten konstant selbst. Hierbei durchlaufen Keratinozyten regulierte Änderungen in der Genexpression sowie der Zellmorphologie. Die Epidermis ist eine wichtige Barriere für den Körper und regelmäßig äußeren Umwelteinflussen wie osmotischen Veränderungen ausgesetzt. Säugerzellen wirken osmotisch bedingtem Anschwellen durch den Prozess der regulatorischen Volumenabnahme entgegen, wobei volumenregulierte Anionenkanäle (VRACs) eine Schlüsselrolle spielen. Die molekulare Identität des VRACs wurde kürzlich aufgeklärt und LRRC8A als essentielle Untereinheit in einem heteromeren LRRC8 Komplex beschrieben. Zudem wurde gezeigt, dass LRRC8 Komplexe die regulatorische Volumenabnahme vermitteln und in verschiedenen Zellen weitere zelltypspezifische Funktionen übernehmen. Allerdings wurde die Rolle von LRRC8A in der Epidermis bisher noch nicht adressiert.

In dieser Studie wurde nun die Funktion von LRRC8A während der hypotonen Stressantwort sowie bei der Differenzierung von Keratinozyten untersucht. Dazu wurde das *LRRC8A* Gen in HaCaT und primären Keratinozyten durch die CRISPR-Cas9 Technologie ausgeschaltet. Durch Messung des Iodideinstroms mittels intrazellulärem Iodidsensor hsYFP konnte gezeigt werden, dass LRRC8A auch in Keratinozyten ein essentieller Bestandteil der volumenregulierten Anionenkanäle ist. Änderungen des Zellvolumens wurden mithilfe des konzentrationssensitiven Fluoreszenzfarbstoffs Calcein verfolgt und es wurde gezeigt, dass LRRC8A an der Zellvolumenabnahme von Keratinozyten beteiligt ist. Zusätzlich wurde gezeigt, dass hypotoner Stress zur Erhöhung der intrazellulären Ca^{2+} Konzentration führt und darüber die VRAC Aktivität und Volumenabnahme verstärkt. Interessanterweise zeigen immunhistochemische Färbungen der humanen Epidermis, dass LRRC8A in der basalen Schicht der Epidermis lokalisiert ist, was auf eine Rolle von LRRC8A beim Wechsel von Proliferation zu Differenzierung hindeutet. Tatsächlich konnte gezeigt werden, dass die Proliferation in Abwesenheit von LRRC8A verringert ist und typische Differenzierungsmarker, nach Induktion der Differenzierung in 2D als auch in rekonstituierten 3D Epidermismodellen, bereits früher gebildet werden. Die Barrierefunktion der rekonstituierten Epidermismodelle, welche durch den Durchtritt des Fluoreszenzfarbstoffes Lucifer yellow gemessen wurde, war durch Ausschalten des *LRRC8A* Gens nicht beeinflusst. Zusammenfassend wurde gezeigt, dass LRRC8A eine wichtige Aufgabe in der hypotonen Stressantwort sowie Proliferation und Differenzierung von humanen Keratinozyten übernimmt. Darüber hinaus lässt sich schlussfolgern, dass LRRC8A ein weiterer Regulator beim Wechsel von Proliferation zu Differenzierung ist und somit zur epidermalen Homöostase beiträgt.

2   Abstract

The human epidermis is the outermost skin layer and constantly self-renewing by proliferation and differentiation of keratinocytes, which undergo regulated changes in gene expression and cell morphology. As important barrier, the epidermis is frequently challenged by various environmental influences including osmotic perturbations. Osmotic cell swelling of all mammalian cells is counteracted by regulatory volume decrease (RVD), which is driven by volume-regulated anion channels (VRACs). Recently, VRACs were identified to be composed of LRRC8 heteromers with LRRC8A as essential VRAC subunit. Furthermore, LRRC8 complexes were shown to mediate RVD and cell type-specific functions in various cell types, whereas its role in human epidermis has not been addressed so far.

Here, the function of LRRC8A during hypotonic stress response of keratinocytes and keratinocyte differentiation was investigated. For this purpose, a *LRRC8A* gene knock-out was generated in HaCaT as well as primary keratinocytes by using the CRISPR-Cas9 technology. It was shown by I^- influx assays using halide-sensitive YFP that LRRC8A is essential for VRAC activity. Cell volume measurements using the concentration-sensitive dye calcein showed that LRRC8A also contributes to regulatory volume decrease. Additionally, hypotonic stimulation of HaCaT cells resulted in an increase of intracellular Ca^{2+} concentration, which enhanced VRAC activity and RVD. Interestingly, immunohistological staining showed preferential localization of LRRC8A in the basal layer of human native epidermis indicating a potential involvement of LRRC8A in keratinocyte transition from proliferation to differentiation. Indeed, it was shown that in the absence of LRRC8A not only proliferation of HaCaT cells was reduced but also expression of differentiation markers occurred earlier after induction of differentiation in 2D as well as in 3D reconstructed HaCaT epidermis equivalents. In contrast, barrier function of HaCaT epidermis equivalents was not altered in the absence of LRRC8A as assessed by penetration of fluorescent dye Lucifer yellow. Taken together, LRRC8A is important in hypotonic stress response as well as keratinocyte proliferation and differentiation. Furthermore, it can be concluded that LRRC8A might be another regulator for the transition from keratinocyte proliferation to differentiation and therefore important for epidermal homeostasis.

3 Introduction

3.1 Structure of the human skin

The skin is the largest human organ involved in a variety of essential functions including body water balance, thermoregulation or immune response. The important barrier between the inner body and the outer environment is composed of three functional and morphological different layers: epidermis, dermis and hypodermis (Figure 3-1) [1].

The hypodermis is the deepest skin layer and an adipose-rich tissue responsible for energy supply, thermal isolation, protection against mechanical shock and protection of underlying structures. It provides a network of nerves, lymph and vascular vessels that are interconnected with the dermis [2, 3].

The dermis sits above the hypodermis. The predominant cell type in the dermis are fibroblasts, which synthesize fibrous and amorphous extracellular matrix proteins thereby forming the connective tissue. Collagen is the mayor dermal protein that provides the skin's flexibility, elasticity and tensile strength. The dermis protects against mechanical injury, has the ability to bind water, regulates body temperature and provides sensory receptors [2]. Furthermore, it contains various skin appendages: apocrine and eccrine sweat glands, hair follicle and sebaceous glands. Lymph and vascular vessels extend from the hypodermis through the dermis up to the dermo-epidermal junction to provide the avascular epidermis with oxygen and nutrients and remove waste products [1].

The epidermis is the outermost skin layer. The stratified epithelium is continuously self-renewing within 28 days. It is composed of keratinocytes, melanocytes, Langerhans cells and Merkel cells. Melanocytes reside in the hair bulb and basal epidermal layer. They form dendrites towards keratinocytes to transfer melanosomes, which are melanin containing granules. After uptake of melanosomes into keratinocytes, melanin colors the skin and protects the cell nucleus from UV light [2, 3]. Langerhans cells originate from the bone marrow and migrate into the skin where they reside as dendritic cells in the suprabasal layer. They function as antigen-presenting cells and recognize, phagocytose, process and present foreign antigens that enter the skin to T-cells and are hence an important part of the immune system forming the immunological skin barrier [2]. Merkel cells differentiate from a keratinocyte progenitor cell *in situ* and are concentrated in the epidermal ridges of developing epidermis. They act as mechanoreceptors and might contribute to the development of skin appendages [4]. Keratinocytes are the pre-dominant cell type in the epidermis with occurrence of 90 – 95 % and progressively differentiate in a basal to suprabasal direction thereby undergoing a variety of morphologic and metabolic changes including arrest of proliferation, increase in size, structural reorganization, synthesis of new proteins and lipids and dehydration [2].

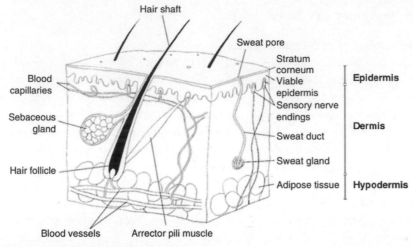

Figure 3-1: Illustration of the human skin

The human skin is composed of three layers: epidermis, dermis and hypodermis. Each layer is composed of different cell types and conducts specific functions. The adipose tissue supplies energy and cushions against physical shock. The dermis is mainly composed of fibroblast, which synthesize proteins that build the connective tissue. Sebaceous glands produce sebum, which is transported to the skin surface where it is involved in skin barrier and antimicrobial functions. Sweat glands produce sweat that is transported through the duct and functions in thermoregulation. Body hair that grows out from the hair follicle is also involved in thermoregulation and touch sensation. Blood vessels provide the skin with oxygen and nutrients. The epidermis is built by keratinocytes that differentiate in a basal to suprabasal direction thereby forming a stratified, self-renewing barrier with important protective function. Adapted from [1].

3.2   Epidermal homeostasis

3.2.1   Proliferation of keratinocytes

The self-renewal capacity of the epidermis and its appendages are based on the existence of different stem cell populations. Keratinocyte stem cells as well as stem cells of hair follicles and sebaceous glands reside in their specific stem cell niches within the epidermis. Although stem cells are already primed as specific progenitors, recent studies showed that their final differentiation is controlled by the microenvironment in the specific niche [5]. Different stem cells still share common features like expression of *keratin 5* and *keratin 14* as well as attachment to the basement membrane and great plasticity during wound healing was shown. This observation further supports the notion that the stem cell fate is influenced by intercellular signaling [5].

The epidermis is able to self-renew due to keratinocyte proliferation in the basal layer and subsequent differentiation in suprabasal direction. A hierarchy of epidermal keratinocytes was found and three subpopulations were defined according to their heterogeneous proliferative potential: keratinocyte stem cells, transient amplifying cells and post-mitotic cells [6]. Keratinocyte stem cells (KSCs) are regarded as stem cell reservoir for ongoing epidermal renewal and have an unlimited self-renewing potential. KSCs can divide symmetrically resulting in two daughter cells that either remain as stem cells or both commit to differentiation. Symmetric cell division that results in two daughter stem cells mainly takes place during embryogenesis [7]. As a consequence cells become crowded, which was recently shown to be a trigger for delamination [8]. During epidermal homeostasis however, most KSC divisions are asymmetric resulting in one daughter cell that remains as stem cell and one daughter cell that commits to differentiation. The division plane can be either perpendicular or parallel to the basement membrane (Figure 3-2). Perpendicular division results in direct separation of daughter cells according to their fate. After parallel division both daughter cells reside (at least for some time) in the basal layer, although one cell resides as stem cell and one as transient amplifying cell [7]. Just recently, it was shown that basal cell division during epidermal homeostasis is a consequence of, rather than the driver for delamination of a neighboring cell [9].Transient amplifying cell (TAs) have a restricted self-renewing potential and undergo a limited number of rapid cell divisions before leaving the cell cycle, delaminating from the basement membrane and committing to terminal differentiation as post-mitotic cells (PMs). The commitment to terminal differentiation is accompanied by morphological changes including increase in cell size and changes in gene expression. The process is tightly controlled and balanced since every corneocyte that sheds from the epidermis has to be replaced by a new cell [10].

Proliferation of basal keratinocytes is induced by insulin-like growth factors (IGFs), fibroblast growth factors (FGFs) and epidermal growth factor receptor (EGFR) ligands that are produced by dermal fibroblasts. Dermis and epidermis are separated by the basement membrane, which is composed of extracellular matrix proteins produced by dermal and epidermal cells. Laminin 5 is the major component of the basement membrane and anchor for basal keratinocytes, which bind to laminin 5 through integrins $\alpha_3\beta_1$ and $\alpha_6\beta_4$ (Figure 3-2). Expression of integrins at the cell surface can be used to isolate the different keratinocyte populations KSCs, TAs and PMs [11]. The switch from proliferation to differentiation involves complex signaling cascades and interplay of various factors to stop cell division and initiate differentiation [12]. *In vitro*, Notch 1 is activated by increased extracellular Ca^{2+} or detachment of murine keratinocytes from the culture plate, which leads to conformational changes of the extracellular domain of Notch 1. This activation results in expression of early differentiation genes and induces expression of the cyclin-dependent kinase inhibitor *p21*, which finally leads to suppressed proliferation [13]. Notch 1-induced expression of *p21* and *p27* can inhibit the transcription factor c-Myc, which was described to induce cell cycle arrest upon pharmacological inhibition [12]. Furthermore, the activated Akt kinase is involved in transition from proliferation to differentiation through phosphorylation of a variety of target molecules, which in a series of signaling pathways induce growth arrest and trigger differentiation [14]. An additional regulator is mTOR, which is downregulated in differentiated keratinocytes and determines fate of keratinocytes to either proliferate or differentiate [15]. The cell cycle is furthermore controlled by the transcription factor p53 whose expression is highest in proliferating keratinocytes and downregulated for terminal differentiation suggesting that p53 coordinates cell cycle and cell division in proliferating keratinocytes [16]. It was found that the cell cycle progresses during differentiation before keratinocytes reside in a special G2/M phase that might be controlled by p53. Although the cell cycle of differentiating keratinocytes progresses and DNA is replicated, cell division does not occur probably due to physical constrains of a rigid cytoskeleton formed by keratin filaments. The process of endoreplication leads to polyploid cells and is proposed as reason for the progressive increase in keratinocyte size and trigger for terminal differentiation [17, 18].

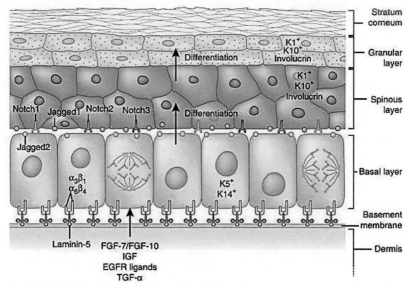

Figure 3-2: Structure and signaling in the epidermis

The epidermis is a stratified epithelium composed of different layers that are build by keratinocytes of different differentiation states. Basal keratinocytes are anchored to the basement membrane by interaction of integrins $\alpha_3\beta_1$ and $\alpha_6\beta_4$ with lamenin 5. Fibroblast-derived growth factors (FGF, IGF, EGFR ligands and TGF-α) influence proliferation of basal keratinocytes. Cell division can occur parallel or perpendicular to the basement membrane. Notch signaling is involved in differentiation and different subsets of keratins (K) are expressed in basal and differentiated keratinocytes. Adapted from [11].

3.2.2 Differentiation of keratinocytes

The differentiation of keratinocytes starts when transient amplifying cells stop proliferating and migrate into the first suprabasal layer as post-mitotic cells. Cornification or keratinization is a special form of cell death and leads to a multilayered epidermis with distinct function and gene expression profile in different layers (Figure 3-2).

The basal layer, also called *stratum basale*, contains mitotically active keratinocytes, which have a flexible cytoskeleton formed by keratin 5 and keratin 14 allowing cell division and migration. Keratins assemble into intermediate filaments thereby contributing to the cytoskeleton of a cell. Type I and II keratin s can be distinguished by being either acidic with low molecular weight (40 – 56 kDa) or basic or neutral with higher molecular weight (52 – 67 kDa) [19]. Keratins are expressed pairwise and one acidic and one basic keratin self-assemble into a central α-helical rod with variable N and C terminal ends. These heterodimers interact with each other forming 10 nm wide filaments [20].

In the *stratum spinosum* expression of keratins changes to keratin 1 and keratin 10, which interact with the existing cytoskeletal network built by keratin 5 and keratin 14 [21]. These larger keratin bundles are linked to desmosomes thereby connecting neighboring cells [22]. Involucrin, envoplakin and periplakin are synthesized in the spinous layer and crosslinked by transglutaminase 1 and transglutaminase 5 to form the scaffold for the cornified envelop just below the cell membrane [23]. Furthermore, members of the S100 protein family are target molecules of transglutaminase 1. These Ca^{2+} binding proteins are localized at the plasma membrane, so it is suggested that S100 proteins mediate Ca^{2+}-dependent signaling during the formation of the cornified envelope [24].

Loricrin represents the major component of the cornified envelope and is expressed in the granular layer. As key structural protein of the cornified envelope loricrin is crosslinked with itself, keratin filaments, filaggrin and small proline-rich proteins by transglutaminase 1, transglutaminase 3 and transglutaminase 5 [25]. Small proline-rich proteins (SPRs) are a family of proteins, which connect loricrin molecules. The ratio between loricrin and SPRs differs between different body sites and determines the rigidity and mobility of the epidermis [26]. The *stratum granulosum* owes its name to keratohyalin granules that are packed with profilaggrin, the precursor of filaggrin [27]. In higher granular layers, profilaggrin is dephosphorylated and proteolysed. The released filaggrin molecules crosslink the keratin intermediate filaments leading to the collapse of keratinocytes into a flat shape [28]. Furthermore, cell organelles including the nucleus are degraded leading to anucleated keratinocytes.

Non-viable and terminally differentiated keratinocytes are called corneocytes, which are assembled in multiple layers to form the *stratum corneum* [22]. These corneocytes are embedded in a specialized lamellar matrix of lipids composed of ceramides, fatty acids and cholesterol, which are released from intracellular lamellar granules into the intercellular space [2, 26]. Associated with the cornified envelope, covalent binding of ceramide to involucrin forms a corneocyte-bound lipid envelope. The lipid envelope might function as scaffold for lamellar lipid organization or cohesion of corneocytes and is therefore important for maintenance of the *stratum corneum* (Figure 3-2) [29].

Although corneocytes are dead cells, the *stratum corneum* is still a biologically living tissue. Enzymes, especially kallikreins, are released from lamellar granules and cleave corneodesmosomes that connect corneocytes and allow cell shedding. The process of desquamation is the final step in the self-renewing process of the epidermis [22].

3.3 Barrier function of the skin

A variety of epidermal barriers are established to collectively prevent the invasion of pathogens, protect against mechanical forces and regulate transepidermal water loss. The skin barrier comprises a microbial, physical, chemical as well as immunological barrier. Healthy skin is colonized by microorganisms, which form an individual- and body site-specific microbiome. Growth of pathogens is directly hindered since the healthy microbiome occupies space and nutrients, produces antimicrobial peptides and inhibits biofilm formation of pathogenic microorganisms [30]. Additionally, humoral and cellular constituents of the immune system are present in the skin contributing to the immunological barrier. Expression of cytokines and chemokines is increased when pathogens infiltrate the skin. Cellular components that are activated upon pathogen entry include Langerhans cells, macrophages, natural killer cells, T-cells or dendritic cells, which reside in and below the epidermis [31]. Acids, hydrolytic enzymes, cytokines and antimicrobial peptides that are produced and distributed in the epidermis are regarded as chemical barrier [32].

The *stratum corneum* together with tight junctions in the *stratum granulosum* compose the physical barrier, which acts as outside-in as well as inside-out barrier restricting the invasion and extrusion of water, solutes and ions. In the *stratum corneum*, corneocytes are organized in a lipid matrix forming an insoluble structure. The lipid-enriched intercellular space counteracts loss of water and solutes from the skin by waterproofing the skin surface [32] (Figure 3-3 A). Furthermore, tight junctions (TJs) control paracellular movement of solutes and ions in the stratum granulosum. Zonula occludens proteins (ZOs), MUPP1, MAGI proteins and cingulin form a cytosolic plaque that is binding to actin and myosin thereby connecting the cytoskeleton to tight junctions. Furthermore, the cytosolic plaque is binding tight junctional transmembrane proteins, which are junctional adhesion molecules (JAMs) and members of the occludin and claudin family. These transmembrane proteins connect adjacent cells of the *stratum granulosum* (Figure 3-3 B) [33]. Disruption of the *stratum corneum* results in increased formation of tight junctions already in the spinous layer as a kind of rescue mechanism [34]. A recent study demonstrated that keratinocytes, which are connected by TJs, adopt a special shape called Kelvin's tetrakaidecahedron. In this shape packaging of cells occurs with minimal free intercellular space, thereby allowing tight cell-cell contact and minimal paracellular passage of water and solutes [35].

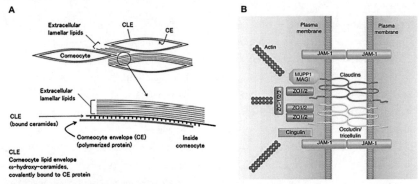

A

B

Figure 3-3: Physical epidermal barrier

The *stratum corneum* and tight junctions establish the physical epidermal barrier. (A) Corneocytes of the *stratum corneum* are surrounded by a rigid cornified envelope (CE). Involucrin, as major protein of the CE, binds ceramides, which form a corneocyte-bound lipid envelope around the CE. The corneocytes are finally embedded in an extracellular matrix of lamellar-arranged lipids to seal the epidermis. Adapted from [36]. (B) Tight junctions are formed in the *stratum granulosum* by a cytosolic plaque of MUPP1, MAGI, ZO1 and 2 and cingulin. The transmembrane proteins JAM1, occludin and the claudin family connect neighboring keratinocytes and restrict paracellular passage of water and solutes. Adapted from [33].

3.3.1 Disturbances of the barrier function

Loss of the entire epidermis can lead to life threatening situations caused by excessive water loss and sepsis. Burns or contact allergens act exogenously, disturb the skin function and cause an immunological reaction. In contrast, endogenous factors like disturbed keratinocyte differentiation or genetic predisposition can trigger inflammation and subsequently lead to perturbed skin barrier [32].

Inflammatory dermatoses can either be caused by or result in a disturbed barrier function. A variety of diseases are linked to mutations in differentiation-associated genes including *transglutaminase 1, loricrin, keratins* or genes involved in lipid synthesis [26]. Furthermore, a disturbed balance between keratinocyte proliferation and differentiation can cause skin disorders. Movement of incompletely differentiated keratinocytes into the *stratum corneum* results in scaling and disturbed desquamation while disturbed barrier function can cause hyperproliferation [37]. A well-known inflammatory skin disease with misbalanced proliferation and differentiation is psoriasis that is characterized by hyperproliferation and abarrant differentiation. Thickening of viable as well as cornified keratinocyte layers manifests clinically as scaly skin lesions at single body sites or extending over larger areas. Psoriatic lesions are infiltrated by immune cells that are stimulated by cytokines [38]. Predisposition involves genetic variation of genes encoding components of the cornified envelop as well as innate and adaptive immune system. However, the disease mostly manifests after external or internal triggers including trauma, skin irritation and microbial infection [39].

3.3.2 Epidermal hydration

The viable part of the epidermis contains a high proportion of water of around 70 % while water content decreases in the *stratum corneum* to 15 – 30 %. This sharp drop is important for separation of viable and non-viable epidermis to retain important solutes in the viable part and induce formation of natural moisturizing factors in the non-viable part [40]. Natural moisturizing factors are hygroscopic substances such as filaggrin fragments, free amino acids and their derivates, sugars and electrolytes. They are distributed in the *stratum corneum* and absorb water from the atmosphere thereby humidifying the outer epidermis [41]. A small fraction of *stratum corneum* hydration depends on the external humidity, while most of the water content in the *stratum corneum* is bound intracellularly and not changing under physiological conditions [42]. Balanced epidermal hydration is important for desquamation; hydrolytic enzymes that are responsible for removal of upper corneocytes require an aqueous environment for proper function. Low water content in the *stratum corneum* leads to impaired desquamation so that corneocytes adhere and accumulate on the epidermal surface resulting in the appearance of dry skin [42].

The *stratum corneum* of the epidermis forms the main physical barrier preventing excessive entry or loss of water. If this skin barrier is disrupted extracellular water can leave the skin thus creating a hyperosmotic milieu. Intracellular water is osmotically driven out of the cell resulting in cell shrinkage, which is actively counteracted by a process called regulatory volume increase. In contrast, regulatory volume decrease occurs after cell swelling induced by extracellular hypotonicity [41]. Hypotonic milieu occurs in the skin when the epidermis is damaged and the skin is exposed to freshwater, which diffuses into the skin [43].

3.4 Mechanism of cell volume regulation

Cell volume regulation does not occur to keep the cell volume at a constant level, but rather acts as a regulator of cellular functions upon endogenous or environmental challenges [44]. Cell volume changes occur during a variety of physiological processes including proliferation, migration, metabolism and cell death as well as during transport and signal processes and can furthermore induce ion channel activity and changes in gene expression [45, 46]. The cell membrane has a limited water permeability, which is increased by water channels called aquaporins [47]. Water is osmotically driven in or out of the cell by changes in protein or ion concentration as well as changes in intra- and extracellular osmolarity. Isosmotic cell volume changes occur in all cells and can be caused by nutrient uptake, activation of ion channels and cell membrane transporters, formation or degradation of osmotically active molecules and degradation of organic substances [48]. Only a few cell types, including cells of the kidney medulla, hepatocytes or epithelial cells are exposed to anisosmotic volume changes that are caused by changes in extracellular osmolarity. Both, isosmotic as well as anisosmotic conditions can cause cell shrinkage or swelling by passive flow of water. Initial cell volume is rapidly re-established by gain or loss of ions and osmolytes through ion channels and transporters building a new osmotic gradient that is followed by water influx or efflux, respectively (Figure 3-4) [49]. A concerted action of a variety of ion channels and transporters is required for fast cell volume regulation perpetuated by a great redundancy in volume-regulating proteins [50].

Cell volume is mainly regulated by anorganic ions that can quickly pass the cell membrane through constantly expressed ion channels. However, large shifts in ion concentration can lead to disturbances in cellular functions. In addition to anorganic ions, organic osmolytes act in volume regulation. Organic osmolytes are 'compatible solutes' that can accumulate in the cytosol up to hundreds of mM without perturbing physiological cell structure or function. Organic osmolytes can be divided into three groups: polyols e.g. sorbitol and myoinositol, amino acids and their derivates e.g. alanine and taurine and methylamines e.g. betaine. Each osmolyte enters the cell by its specific transporter whose expression is adapted by cell volume changes [51, 52]. However, osmolyte uptake does only slightly contribute to regulatory cell volume increase due to slow influx rate. Regulatory volume decrease on the other hand is significantly influenced by efflux of multiple osmolytes that seems to occur through only one type of channels; volume-regulated anion channels (VRACs), which are differently composed of LRRC8 proteins [50, 53].

Anisosmotic shrinkage occurs when cells are exposed to a hyperosmotic extracellular fluid, which leads to efflux of water out of the cell. Initial cell volume is restored in a process called regulatory volume increase (RVI) where a new concentration gradient of Na^+, K^+ and Cl^- is established thus leading to osmotically driven influx of water into the cell. Increase in intracellular Na^+ and Cl^- concentration is mediated by Na^+-H^+ (NHE1) and Cl^--HCO_3^- (AE2) exchanger as well as Na^+-K^+-$2Cl^-$ cotransporters (NKCC1 and 2). Efflux of H^+ through NHE1 leads to alkalization of the cytosol, which is counteracted by efflux of HCO_3^- through activated AE2 resulting in electroneutral influx of Na^+ and Cl^- with only minor changes of intracellular pH [50, 54]. In addition, expression of organic osmolyte transporters is upregulated leading to an increased influx of organic osmolytes mainly taurine that leads to long-term adjustment of cell volume [50, 55].

On the contrary, exposure of cells to hypotonic solution leads to anisosmotic cell swelling since water is osmotically driven into the cell. During regulatory volume decrease K^+ and Cl^- is transported out of the cell to establish a new concentration gradient that is followed by water to restore initial cell volume. Besides electroneutral release of K^+ and Cl^- via

K^+-Cl^--cotransporters (KCC1-4) and concerted activity of K^+-H^+-exchanger and Cl^--HCO_3^- exchanger, K^+ and Cl^- ions leave the cell through distinct channels. Upon cell swelling K^+ conductance is increased leading to outward directed K^+ flow thereby hyperpolarizing the cell, which in turn induces efflux of Cl^-. Various swelling activated K^+ channels have been described in different cell types including stretch-activated K^+ channels, voltage-dependent K^+ channels and Ca^{2+}-activated channels of small, intermediate and big conductance. There is no ubiquitary K^+ channel described for all investigated cell types. In contrast, remarkably constant Cl^- currents were measured, which are solely activated upon cell swelling and mediated by volume-regulated anion channels (VRACs). Additionally, in some cell types Ca^{2+}-activated Cl^- channels might be activated that require increase of the intracellular Ca^{2+} concentration and which can be distinguished from VRACs by their electrophysiological properties [49, 50, 56].

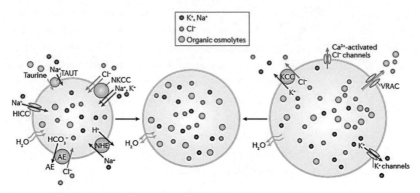

Figure 3-4: Ion transport pathways involved in cell volume regulation
Cell volume is regulated by changes of intracellular Na^+, K^+ and Cl^- concentration, which drive water osmotically over the membrane to adjust the physiological cell volume or restore the initial cell volume after perturbations. To increase cell volume Na^+ and Cl^- are transported into the cell by various transporter systems to establish an osmotic gradient resulting in influx of water. In contrary, after cell swelling K^+ and Cl^- are released from the cell by transporters and ion channels to osmotically drive water out of the cell resulting in reduction of cell volume. Modified from [50].

Cell swelling leads to increase of the intracellular Ca^{2+} concentration by influx of Ca^{2+} from the extracellular space and/or release from intracellular stores in various, but not all, investigated cell types. The pathways that lead to increased Ca^{2+} concentration seem to be variable and include stretch-activated ion channels, Na^+-Ca^{2+} exchange, inositoltrisphosphate-dependent pathways, Ca^{2+}-induced Ca^{2+} release and store-operated Ca^{2+} entry [57]. The role of Ca^{2+} in RVD is cell type-dependent since some cell types require extracellular Ca^{2+} for activation of RVD, other cells types depend on Ca^{2+} release from intracellular stores and some cells show only weak to no Ca^{2+} dependency [58].

In cultured human keratinocytes, hypotonic stimulation results in increase of the intracellular Ca^{2+} concentration that depends on the presence of extracellular Ca^{2+} and purinergic signaling. However, increase of intracellular Ca^{2+} concentration is not sufficient to activate VRACs or mediate RVD. Although human keratinocytes react to hypotonic stimulation with chloride efflux, magnitudes and kinetics of RVD differed between the reports [59-61]. A systematic study on the effect of Ca^{2+} on VRAC activity and RVD of keratinocytes is still missing.

3.4.1 Cell volume regulation during proliferation

As mentioned above, proliferation and differentiation of keratinocytes are essential for epidermal homeostasis. Although not investigated in keratinocytes, cell volume changes are relevant for cell proliferation since cell size is one criteria on the long list of checkpoints during cell cycle progression. Mitosis can only occur when cell volume condensates to a cell type-specific size in a process termed pre-mitotic condensation that is mediated by solute and water efflux as seen during RVD [62]. The accompanying mitotic cell rounding is actively balanced by actomyosin contraction and intracellular osmotic pressure [63]. Disbalance leads to altered cell volume and might finally result in incorrect cell division due to changes in intracellular protein or solute concentrations or spacial rearrangements [64]. After mitotic division, the cell volume is at its minimum and increases during G1 phase. Entry into S phase is again determined by cell size with smaller cells entering S phase faster than larger cells [65]. At the end of G2 phase the Na^{+}-H^{+} exchanger NHE1 is activated, thereby increasing cell volume and intracellular pH to allow the transition from G2 to M phase [66]. Critical cell volume regulators are K^{+} and Cl^{-} channels for which a cell cycle dependent activation was found. K^{+} and Cl^{-} permeability is lowest in S phase and highest in G1 phase. Cl^{-} currents, that are closely associated with RVD, as well as K^{+} currents inversely correlated with cell volume [67, 68].

3.4.2 Volume-regulated anion channels

Although the molecular identity of chloride channels involved in RVD has long been unknown, they were deeply investigated as volume-regulated anion channels (VRAC). VRAC-mediated chloride currents are activated upon hypotonic stimulation with modest outward rectification that is inactivated at positive membrane potential [69]. VRACs display a weak anion selectivity according to Eisenman's sequence I ($SCN^{-} > I^{-} > NO3^{-} > Br^{-} > Cl^{-} >$ glycine $> F^{-} >$ taurine $>$ glutamate $>$ aspartate). They are not only permeable for small anions, but also for bigger anionic molecules like ATP [70] and taurine, which is also involved in regulation of cell volume [69, 71].

Activation of VRAC channels seems to include multiple factors. Cell swelling is the most obvious one that results in various cellular alterations including changes in membrane tension, curvature and arrangement of membrane proteins as well as changes in intracellular ion and protein concentration [72]. Different studies showed that VRAC activation requires intracellular ATP. It was postulated that not phosphorylation but rather non-hydrolytic ATP binding is involved in activation of VRAC. However, VRAC activation solely depends on ATP, when cells swell only slightly and ATP dependency is overruled by stronger cell swelling [70, 73]. Besides hypotonic stimulation VRACs can be isovolumetrically activated by a variety of stimuli including induction of intracellular GTPγS and reduction in intracellular ionic strength. It was discussed that cytoplasmic ionic strength defines the volume set point of a cell. VRACs are activated by reduction in cytoplasmic ionic strength even without cell swelling but also when cells swell at constant ionic strength [73]. Furthermore, VRACs can be activated by stimulation of purinergic and bradykinin receptors, which both lead to Ca^{2+} signaling [74]. However, VRACs per se are activated independently from intracellular Ca^{2+} concentration. VRAC currents can be blocked by various known anion channel inhibitors like NPPB, DIDS, NFA and DCPIB. However, most inhibitors are not exclusively blocking VRACs, but also other anion channels as well as cation channels, thereby hindering the identification of the molecular identity of VRACs [49, 75, 76].

Multiple proteins were proposed as potential VRAC molecules. First candidates such as P-glycoprotein, pICln, ClC-3, band three anion exchanger and phospholemman are no longer considered since their biophysical properties do not completely correlate to those observed for VRACs [75, 76]. Still under debate are bestrophins, the TMEM16 protein family, the LRRC8 protein family [77] and the TTYH protein family [78, 79]. Bestrophin1 was first proposed as VRAC in *Drosophila* [80] and more recently in mouse spermatozoa and human retinal pigment epithelium cells [81]. However, measurement of volume-activated currents of mast cells harvested from best1/best2 null mice [82], identification of pentameric bestrophin channels as Ca^{2+}-activated Cl^- channels [83] and the restricted tissue expression in humans argue against the role of bestrophin 1 as VRAC. The human TMEM16 protein family consists of ten proteins, whose expression and functionalities are highly diverse in different tissues although all family members are expected to form channels [84, 85]. TMEM16A is activated upon cell swelling and knock-down of TMEM16A, F, H and J reduced swelling-induced chloride conductance [86]. However, TMEM16A displays typical currents of Calcium-activated chloride channels ($I_{Cl,(Ca)}$) and Ca^{2+}-dependent activation and thus is questioned to be a VRAC. The *Drosophila melanogaster* tweety homologue family is the latest proposed VRAC candidate. *Tweety* genes were first described in *Drosophila melanogaster* before three human homologues (TTYH1-3) were identified [87]. It is suggested that TTYH family members assemble into homomers to form the channel pore [79]. They were initially described to serve as maxi-Cl^- conducting channels that are swelling-activated (TTYH1) or activated by increase of intracellular Ca^{2+} (TTYH2 and 3) [87]. Newer studies are challenging the characterization of differently activated maxi-Cl^- conducting channels since in gastric cancer cells

TTYH1 and 2 [78] and in astrocytes all three family members [79] mediate typical swelling-induced VRAC currents.

3.4.3 The LRRC8 family

LRRC8A, also known as SWELL1, was identified as VRAC candidate, which was the first to fulfill all known physiological properties of VRACs [88, 89]. An ubiquitous expression of *LRRC8A* and the four other family members (*LRRC8B-E*) was proposed by phylogenetic analysis [90] and expression of *LRRC8A* was confirmed in a variety of cell lines and tissues including HEK293 [89], HeLa [88], immune cells [88, 91], cancer cell lines [89, 92, 93], adipocytes [94], pancreatic β-cells [95] and neuronal cells [96, 97]. LRRC8A is composed of four transmembrane domains and an intracellular domain of 16 leucine-rich repeats (LRR) (Figure 3-5). A similar topology is proposed for LRRC8B to E with little alterations in number of LRR domains [98] and positions for post-translational modifications. Formation of a hexameric channel is proposed based on the sequence similarity with pannexins [90].

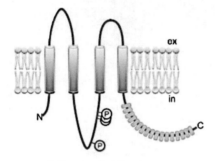

Figure 3-5: Topology of LRRC8A

LRRC8A contains four transmembrane domains and 16 leucine-rich repeats with intracellular N- and C-termini. Topology was initially described by two groups independently [88, 89]. Possible phosphorylation sites are depicted in the figure modified from [89].

Meanwhile, hexamer formation of LRRC8A was confirmed by cryo electron microscopy (Figure 3-6). LRRC8A hexamers assemble into a jellyfish-like shape with a constriction of the pore towards the extracellular site and a horseshoe shape of the cytosolic LRR domains. The six LRRC8A monomers assemble symmetrically around a central axis perpendicular to the membrane with a C6 symmetry of the transmembrane domains. When LRRC8A-monohexamers were detergent-stabilized, the LRR domains arrange in a C3 symmetry forming a trimer of dimers [99-101], but a C6 symmetry when reconstructed in lipid nanodiscs [102]. The LRR domain is flexible and proposed to be involved in channel gating. The ion-conducting pore is formed by the extracellular loop, the transmembrane domains and the intracellular linker domains, which are wide enough to allow passage of anions and osmolytes [99-102].

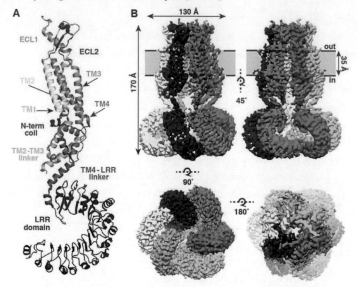

Figure 3-6: Structure of a LRRC8A monomer and assembly into hexamers

(A) Structure of a LRRC8A monomer depicting the tertiary structure of extracellular loops (ECL), transmembrane domains (TM) and leucine-rich repeats (LRR). (B) Cryo electron microscopy revealed the assembly of LRRC8A into hexamers. The channel is depicted from membrane plane, cytosolic and extracellular site showing C6 symmetry of the transmembrane domain and C3 symmetry of the LRR domain. Adapted from [100]

LRRC8A is essential for VRAC activity since knock-out resulted in abolishment of chloride efflux from swollen cells. However, at least one other family member is necessary for mediating swelling-induced chloride currents ($I_{Cl,swell}$) [88, 89]. LRRC8A does not only assemble with one but multiple other subunits, which gives rise to a variety of diversely composed VRAC channels that can be expressed in the same cell. LRRC8 subunit-specific efflux of organic osmolytes and neurotransmitters was reported, which are all known to be mediated by VRACs in addition to chloride. LRRC8D-containing VRACs allow efflux of uncharged molecules like taurine, myo-inositole or GABA while VRACs containing LRRC8C and LRRC8E mediate efflux of charged molecule including glutamate, lysine and aspartate [53, 103]. Furthermore, VRAC inactivation kinetics [89], single-channel conductance [104] and selectivity for cisplatin and taurine uptake [92] are determined by LRRC8 subunit composition.

Activation of VRACs was studied even before the molecular identity was known as described above. However, many aspects that were described to activate VRACs were not proven after the molecular identification and still very little is known about the activation of LRRC8A. Reduction of cytoplasmic ionic strength was enough to activate LRRC8A in reconstituted bilayer system [104] which could not be confirmed in another study [105]. Structural analysis showed that the C-terminal LRR domains rearrange upon hypotonic stimulation thereby probably influencing channel gating. Furthermore, several putative phosphorylation sites within LRRC8A are predicted and activation of protein kinase D is important for LRR domains rearrangement as well as VRAC currents [105].

In addition to its function in mediating chloride and osmolyte efflux during regulatory volume decrease, LRRC8A is associated with a variety of physiological functions. LRRC8A was first isolated from a patient suffering from congenital agammaglobulinemia, a disease that is associated with defective B cells development. Truncation of LRRC8A in these B cells led to the suggestion that LRRC8A is involved in B cell maturation [98]. However, LRRC8A is also important for normal physiology of various tissues as *LRRC8A*$^{-/-}$ mice show a severely altered phenotype including increased postnatal lethality, growth retardation, curly hair, hind limb weakness, thin skeletal muscle bundles, progressive hydronephrosis (vacuolation of renal tubular cells), sterility, smaller thymus and epidermal hyperkeratosis (thickening of *stratum corneum* and *stratum granulosum* with abnormal quantities of keratins) [106]. A milder phenotype was observed in mice with truncation of the terminal 15 LLRs in LRRC8A, which still conduct little VRAC currents [107]. In addition, LRRC8A is important for glucose metabolism and insulin signaling in adipocytes [94] as well as pancreatic β cells [108] and sperm development in mice [109, 110]. Since LRRC8A-containing channels also release excitatory amino acids, the LRRC8 family plays a role in neuronal communication [97, 103]. LRRC8A seems to be important for proliferation as siRNA mediated knock-down of LRRC8A suppressed activation of PI3K/Akt resulting in inhibition of angiotensin II-induced proliferation of cerebrovascular smooth muscle cells [111]. This finding is supported by the fact that reduced proliferation of glioblastoma cells also correlates with downregulation of LRRC8A [112]. In conclusion, LRRC8A has an impact in a great variety of tissues, which underpins the essential role of LRRC8A in tissue homeostasis throughout the body.

4   Aim of the study

Mammalian cells counteract osmotic cell swelling by regulatory volume decrease (RVD) including activation of volume-regulated anion channels (VRACs). VRACs were identified as LRRC8 heteromers and LRRC8A was shown to be essential for VRAC activity and RVD in different cell types, however, keratinocytes were not addressed so far. Since the epidermis is constantly challenged by environmental perturbations including osmotic changes, it can be hypothesized that LRRC8A is also involved in hypotonic stress response of keratinocytes. Thus, expression of *LRRC8A* will be analyzed in native human epidermis as well as in cultured primary keratinocytes and immortalized HaCaT cells. To determine the contribution of LRRC8A to hypotonic stress response, cultured keratinocytes devoid of *LRRC8A* will be generated by CRISPR-Cas9 technology and used to measure VRAC activity as well as RVD in the absence of LRRC8A.

Osmotically-induced keratinocyte swelling, which might be counteracted by activation of LRRC8A, occurs naturally only after barrier disruption, which can arise in irritated or diseased skin. In addition, LRRC8A-mediated cell volume regulation is an integral part of many general cellular processes and associated with cell type-specific functions. Since keratinocytes undergo endogenous changes in cell morphology during epidermal homeostasis the involvement of LRRC8A in homeostatic proliferation and differentiation of keratinocytes will be analyzed. Finally, contribution of LRRC8A to epidermal barrier function will be assessed. In summary, results from this study will contribute to understand the role of LRRC8A in hypotonic stress response of keratinocytes as well as in epidermal homeostasis. Moreover, it will serve as a starting point to study LRRC8A in skin diseases associated with disturbed skin barrier, which are known to aggravate upon hypotonic stress.

5   Results

5.1   LRRC8A is expressed in basal keratinocytes of human epidermis and in cultured human keratinocytes

Expression of *LRRC8A* was shown in several tissues and cell lines [88, 113] but notably not in native human skin, which was the first question to answer in this study. Therefore, punch biopsies were taken from skin of four healthy donors and processed by standard immunohistological procedures. LRRC8A was detected with a LRRC8A-specific primary antibody and secondary antibody coupled to alkaline phosphatase. Antibody binding was visualized by an alkaline phosphatase catalyzed enzymatic reaction resulting in red staining.

In all of the four tested donors, LRRC8A antibody staining in the epidermis was preferentially detected in basal keratinocytes, while less LRRC8A antibody staining was observed in upper, further differentiated layers (Figure 5-1 A and Supplement Figure S1). As a control for antibody specificity, the immunohistologic procedure was also performed using an isotype control antibody instead of LRRC8A antibody with the result that no staining was detected (Figure 5-1 B) supporting the specificity of the shown staining.

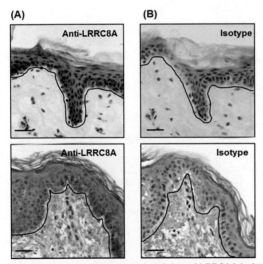

Figure 5-1: Immunohistological staining of LRRC8A in human skin biopsies

Punch biopsies from two healthy donors were taken and sections were subjected to immunohistological staining with Anti-LRRC8A antibody (A) or isotype control antibody (B) and nuclei staining with hematoxylin (blue). LRRC8A antibody binding is visualized by red staining and shows preferential expression of *LRRC8A* in the lowest epidermal layer, the *stratum basale*. Black line illustrates the border between dermis and epidermis. Bars represent 30 μm.

After LRRC8A has been detected in native human epidermis, it was asked whether *LRRC8A* and the other four members of the *LRRC8* gene family, which are important for the function of LRRC8 complexes, are also expressed in cultured keratinocytes. Therefore, total mRNA from keratinocyte cell line HaCaT and primary keratinocytes NHEK was subjected to RNA sequencing. In addition, HEK293 cells were included as a reference cell line since LRRC8A was first identified as VRAC component in HEK293 cells [88, 89]. Expression pattern of all five family members (*LRRC8A-E*) was similar in HaCaT and NHEK cells (Figure 5-2). Highest mRNA abundance was detected for *LRRC8A* (45 FPKM) followed by *LRRC8D, LRRC8C, LRRC8E* and *LRRC8B* (5 FPKM). In contrast to keratinocytes, abundance of all five *LRRC8* mRNA transcripts was less different among each other in HEK293 cells (FPKM around 5). Interestingly, mRNA abundance of *LRRC8A* was around seven times higher in keratinocytes compared to HEK293 cells. Thus, this cell type-specific expression pattern with *LRRC8A* showing highest mRNA abundance as well as the localization of LRRC8A in basal keratinocytes of native epidermis raise the question what particular function LRRC8A exerts in the epidermis.

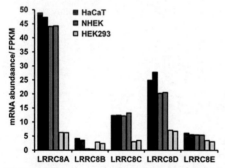

Figure 5-2: Expression of the LRRC8 gene family in human keratinocytes and HEK293 cells
Total mRNA isolated from HaCaT, NHEK and HEK293 cells was used for RNA sequencing to determine mRNA transcript abundances of *LRRC8A-E*. Transcript abundances are displayed in fragments per kilobase of exon per million fragments mapped (FPKM) from two independent RNA preparations per cell line.

5.2 Keratinocytes respond to hypotonic stress with activation of VRACs during regulatory volume decrease

After having shown that *LRRC8A* is expressed in HaCaT and NHEK cells, its functionality in hypotonic stress response was investigated in these cells. In general, hypotonic stimulation leads to cell volume increase, which is counteracted by RVD involving activation of VRACs [49]. First, the extend of VRAC activation and its contribution to RVD in human keratinocytes was addressed.

5.2.1 Measurement of VRAC activity by hypotonicity-induced hsYFP quenching

To determine VRAC activity, I⁻-dependent quenching of the halide-sensitive yellow fluorescent protein (hsYFP) was measured. Activated VRACs not only allow Cl⁻ efflux but are also permeable for other ions. I⁻ enters the cell along a concentration gradient and selectively binds to hsYFP resulting in quenching of intracellular hsYFP fluorescence that is measured to indirectly determine VRAC activity in real-time [114].

In order to measure VRAC activity in HaCaT cells, first of all, these cells were genetically engineered to stably express the iodide sensor hsYFP. HaCaT cells were transduced with a lentivirus to integrate a gene expression cassette encoding for hsYFP and puromycin-N-acetyltransfease to generate a resistance against puromycin. After antibiotic selection and single cell cloning, a clonal HaCaT cell line with stable and robust expression of *hsYFP* (HaCaT-*hsYFP*) was raised and used for further analysis.

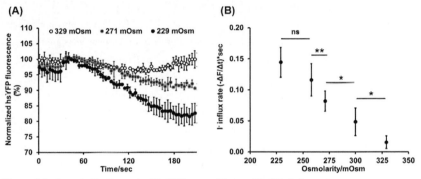

Figure 5-3: Osmolarity-dependent hsYFP quenching and iodide influx rate

(A) Yellow fluorescence emitted from HaCaT-*hsYFP* cells was recorded, normalized to fluorescence at the time point of stimulation and plotted over time. Cells were stimulated with isotonic buffer (329 mOsm) as control. Stimulation with hypotonic buffer of different osmolarity (271 mOsm and 229 mOsm) resulted in osmolarity-dependent decrease of hsYFP fluorescence. One representative experiment is shown depicting the mean with SD from three technical replicates. (B) I⁻ influx rate (-ΔF/Δt) was calculated from hsYFP fluorescence decrease over time. HaCaT-*hsYFP* cells were stimulated with hypotonic buffers of decreasing osmolarity showing osmolarity-dependent quenching of hsYFP. Graph shows the mean with SD of 17 measurements from two independent experiments.

To proof that HaCaT-*hsYFP* cells can be used to measure hypotonicity-induced hsYFP quenching, HaCaT-*hsYFP* cells were stimulated with hypotonic buffers of different osmolarity. HsYFP fluorescence was recorded in real-time, normalized to hsYFP fluorescence at the time point of stimulation and plotted over time. HsYFP quenching decreased by 20 % within 150 sec upon strong hypotonic stimulation (229 mOsm) (Figure 5-3 A). Stimulation with moderate hypotonicity (271 mOsm) resulted in only 10 % hsYFP quenching within the same time. As control, isotonic buffer (329 mOsm) was applied and hsYFP fluorescence decreased by 4 %. To quantify hypotonicity-induced hsYFP quenching the iodide influx rate ($-\Delta F/\Delta t$) was calculated from hsYFP fluorescence decrease (F) over time (t). Importantly, by increasing osmolarity the I⁻ influx rate decreased (Figure 5-3 B) demonstrating that hsYFP quenching is osmolarity-dependent. For further experiments, an osmolarity of 229 mOsm was used to induce maximal hsYFP quenching.

To test whether VRACs are the major chloride channels mediating hypotonicity-induced hsYFP quenching, the effect of pharmacological inhibitors against different types of chloride channels was investigated. Incubation of HaCaT-*hsYFP* cells with VRAC inhibitor carbenoxolone (CBX) [88] reduced hypotonicity-induced hsYFP quenching to 6 % compared to 21 % quenching upon hypotonic stimulation without inhibitor (Figure 5-4 A). Additionally, pretreatment of HaCaT-*hsYFP* cells with VRAC inhibitor DCPIB [115] led to 8 % reduction in hsYFP quenching although hsYFP fluorescence first increased by 4 % upon addition of hypotonic buffer (Figure 5-4 B).

To determine the efficiency of CBX and DCPIB, HaCaT-*hsYFP* cells were incubated with increasing inhibitor concentrations. Normalized iodide influx rates were plotted over inhibitor concentration and concentration of half-maximal inhibition was determined by curve fitting. Inhibition was dose-dependent with an $IC_{50}(CBX)$ = 36.9 ± 7.8 µM and $IC_{50}(DCPIB)$ = 43.9 ± 7.2 µM (Figure 5-4 C and D). The highest applicable concentration of CBX or DCPIB achieved 90 % or 75 % inhibition of I⁻ influx rate (Figure 5-4 E). DIDS inhibits VRACs as well as other anion channels [61] and yielded 50 % reduction of I⁻ influx rate at the highest applicable concentration. In contrast, niflumic acid (NFA), which inhibits calcium-activated chloride channels (CaCCs) [116], did not inhibit hypotonicity-induced I⁻ influx (Figure 5-4 E). Taken together, it was shown that VRAC inhibitors CBX and DCPIB, but not CaCC inhibitor NFA, blocked hypotonicity-induced I⁻ influx thereby confirming that hypotonicity-induced hsYFP quenching by I⁻ influx is mediated by VRACs.

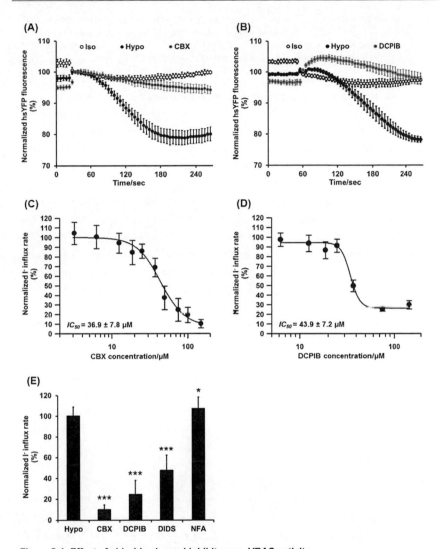

Figure 5-4: Effect of chloride channel inhibitors on VRAC activity

(A, B) HaCaT-*hsYFP* cells were incubated with 150 µM CBX (A) or 100 µM DCPIB (B) prior to isotonic (Iso, 329 mOsm) or hypotonic (Hypo, 229 mOsm) stimulation. HsYFP fluorescence was normalized to fluorescence at the time point of stimulation and plotted over time. Reduction of VRAC activity by CBX and DCPIB is shown. One representative experiment is presented in A and B depicting the mean with SD from six technical replicates. (C, D) HaCaT-*hsYFP* cells were incubated with increasing concentrations of CBX (C) or DCPIB (D). I⁻ influx rate was calculated from hsYFP quenching over time and I⁻ influx rate of inhibitor-treated cells was related to I⁻ influx rate after hypotonic stimulation without inhibitor. Normalized I⁻ influx rate was plotted over inhibitor concentration and *IC₅₀* values were determined from curve fitting. One representative curve for each inhibitor is shown. ----- continued on the next page -----

(E) HaCaT-*hsYFP* cells were pretreated with 150 µM CBX, 100 µM DCPIB, 50 µM DIDS or 50 µM
NFA and normalized I⁻ influx rate was calculated to assess the inhibitory effect. Bars represent the
mean with SD of at least 16 measurements from three independent experiments. Asterisks depict
the statistical significance compared to hypotonic stimulation (Hypo).

5.2.2 Measurement of regulatory volume decrease using calcein

Based on the osmolarity-dependency of hsYFP quenching, which was prevented by
VRAC inhibitors, it was shown above that VRACs are the major channels activated upon
hypotonic stimulation in HaCaT cells. Thus, it was subsequently investigated whether
HaCaT cells respond to hypotonic stress with cell volume increase and regulation of cell
volume by RVD.

To measure cell volume changes, cells were loaded with the non-fluorescent and
membrane-permeable molecule calcein-AM. Intracellular esterases cleave off the
acetoxymethyl group releasing the non-permeable and green fluorophore calcein, whose
fluorescence intensity depends on the calcein concentration and therefore on cell volume.
Cell swelling leads to increase of calcein fluorescence and decreased fluorescence
indicates cell shrinkage by RVD. Therefore, this method allows indirect monitoring of cell
volume changes in real-time [117].

To proof that calcein is suitable to measure cell volume changes in response to hypotonic
stimulation, HaCaT cells were stimulated with buffers of different osmolarity after
loading with calcein-AM. Calcein fluorescence was recorded, normalized to fluorescence
at the time point of stimulation and plotted over time. Upon strong hypotonic stimulation
(150 mOsm) calcein fluorescence rapidly increased reaching a maximum at 115 % and
decreased to 104 % within 30 sec indicating swelling and shrinkage of HaCaT cells.
Moderate hypotonic stimulation (236 mOsm) induced increase of calcein fluorescence to
107 % and decrease to baseline level indicating less cell swelling and slower RVD. Control
stimulation with isotonic buffer (329 mOsm) immediately resulted in a 3 % drop of calcein
fluorescence, which remained stable throughout the measurement (Figure 5-5 A).

The difference between maximal calcein fluorescence and baseline calcein fluorescence (F)
over time (t) was used as a quantitative measure of the speed of RVD (-$\Delta F/\Delta t$).
Quantification showed that stimulation with higher osmolarities resulted in slower RVD
(Figure 5-5 B). In summary, changes in calcein fluorescence depend on the strength of
hypotonic stimulation and stronger osmotic stimulation resulted in faster cell volume
decrease of HaCaT cells. For further experiments, a final osmolarity of 150 mOsm was
used to induce maximal cell volume changes.

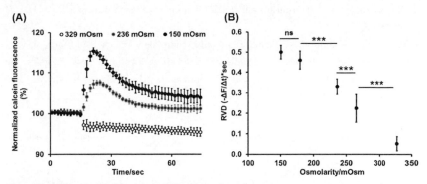

Figure 5-5: Osmolarity-dependent cell volume changes

HaCaT keratinocytes were loaded with calcein-AM prior to measurement. (A) Change in calcein fluorescence upon isotonic (329 mOsm) or hypotonic stimulation (236 mOsm and 150 mOsm) were recorded, normalized to fluorescence at the time point of stimulation and plotted over time. Calcein fluorescence increase and decrease indicate cell volume increase and cell volume decrease. One representative experiment is presented depicting the mean with SD from six technical replicates. (B) The fluorescence decrease over time ($-\Delta F/\Delta t$) after maximal cell swelling was used as a measure of the speed of regulatory volume decrease (RVD), which is osmolarity-dependent. Graph shows the mean with SD of 15 measurements from two independent experiments.

5.2.3 Contribution of VRACs to regulatory volume decrease

Hypotonic stimulation of HaCaT keratinocytes leads to cell volume increase followed by cell volume decrease by RVD. Other cell types were used to show that VRACs are the key players in RVD [49]. Thus, the question arose whether VRACs mediate RVD also in keratinocytes.

To answer this question, HaCaT keratinocytes were incubated with different known chloride channel inhibitors before measuring hypotonicity-induced RVD using calcein. Hypotonic stimulation of CBX-incubated HaCaT cells resulted in 4 % calcein fluorescence increase that was lower than without CBX incubation (7 %). Additionally, calcein fluorescence decreased more slowly after CBX incubation and did not reach baseline level (Figure 5-6 A) showing that CBX inhibits cell swelling as well as cell shrinkage. Pretreatment of HaCaT cells with DCPIB resulted in 5 % increase of calcein fluorescence and slower and incomplete fluorescence decrease compared to 7 % increase and fast fluorescence decrease upon hypotonic stimulation without DCPIB (Figure 5-6 B) showing that DCPIB reduced cell volume increase and decrease.

To quantify these inhibitory effects, RVD of pretreated cells was calculated and related to RVD of HaCaT cells upon hypotonic stimulation without inhibitor. At the highest applicable concentration CBX showed the highest potency with 58 % reduction of RVD (Figure 5-6 C). DCPIB and DIDS inhibited RVD by 32 % and 39 %. The CaCC inhibitor

NFA did not inhibit RVD indicating that CaCCs are not activated during RVD of HaCaT keratinocytes.

None of the tested VRAC inhibitors could completely block RVD and HaCaT keratinocytes were still able to partially reduce their cell volume. This suggests that besides VRACs, additional channels and transporters are contributing to hypotonic stress response.

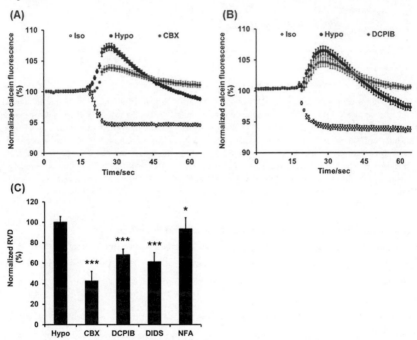

Figure 5-6: Effect of chloride channel inhibitors on RVD of HaCaT cells

HaCaT cells were loaded with volume-sensitive dye calcein-AM and incubated with 150 µM CBX (A) or 100 µM DCPIB (B) prior to stimulation. (A, B) Calcein fluorescence upon isotonic (329 mOsm) or hypotonic (150 mOsm) stimulation was normalized to fluorescence at the time point of stimulation and plotted over time. One representative experiment for CBX (A) and DCPIB (B) is shown depicting the mean with SD from five technical replicates. (C) RVD after inhibitor incubation was calculated from calcein fluorescence decrease over time and related to RVD upon hypotonic stimulation without inhibitor (Hypo). Calcein loaded HaCaT cells were pretreated with CBX (150 µM), DCPIB (100 µM) and DIDS (50 µM) or NFA (50 µM). Bars represent the mean with SD of at least 22 measurements from three independent experiments. Asterisks depict the statistical significance compared to hypotonic stimulation (Hypo).

5.3 Hypotonic stimulation leads to elevation of intracellular Ca^{2+} which increases VRAC activity and RVD

Different studies show that VRACs do not require Ca^{2+} although a supportive role of Ca^{2+} for activation of VRACs and during RVD is frequently discussed [61, 118-121]. Gönczi *et al.* found that HaCaT cells respond to hypotonic stimulation with an increase in intracellular Ca^{2+} concentration and that chelation of intracellular Ca^{2+} did not prevent hypotonicity-induced Cl^- currents [60]. However, it is yet unknown whether the intracellular Ca^{2+} increase affects VRAC activity and if Ca^{2+} is involved in RVD of HaCaT keratinocytes.

5.3.1 Hypotonic stimulation increases intracellular Ca^{2+} concentration

To determine changes in intracellular Ca^{2+} concentration, HaCaT cells were loaded with Fluo4-AM as a widely used Ca^{2+} indicator [122]. Intracellular esterases cleave off the acetoxymethyl (AM) group from Fluo4-AM thereby releasing Fluo4 whose fluorescence increases upon binding of Ca^{2+}.

HaCaT cells were stimulated and Fluo4 fluorescence was recorded, normalized to fluorescence at the time point of stimulation and plotted over time. Hypotonic stimulation with buffer containing 2 mM $CaCl_2$ resulted in a fast 36 % increase of Fluo4 fluorescence indicating an increase of the intracellular Ca^{2+} concentration. Within 10 sec Fluo4 fluorescence decreased by one third where a short plateau was established before fluorescence further decreased to the level of isotonic control (Figure 5-7 A).

To determine the origin of intracellular Ca^{2+}, hypotonic buffer without $CaCl_2$ was used for stimulation, which resulted in 20 % increase of Fluo4 fluorescence, which was lower compared to hypotonic stimulation with 2 mM $CaCl_2$. Fluorescence decreased similar to stimulation with buffer containing 2 mM $CaCl_2$; showing a first fast decrease, short plateau phase and final decrease to isotonic control. To chelate residual traces of Ca^{2+}, 1 mM EGTA was added to the hypotonic buffer without $CaCl_2$ and stimulation resulted in only 1 % increase in Fluo4 fluorescence without further changes. Isotonic stimulation containing 2 mM $CaCl_2$ was used as control and Fluo4 fluorescence increased by 5 % until the end of the measurement (Figure 5-7 A). A similar increase in Fluo4 fluorescence was recorded for isotonic buffer without $CaCl_2$ as well as after addition of EGTA. These curves are not shown in Figure 5-7 A to facilitate visualization.

Maximal fluorescence increase ($\Delta F=F_{max}-F_0$) related to baseline fluorescence (F_0) was used as a measure of changes of intracellular Ca^{2+} concentration ($\Delta F/F_0$). In the presence of 2 mM extracellular Ca^{2+} hypotonic stress led to a strong increase of intracellular Ca^{2+}, which was reduced when extracellular Ca^{2+} was omitted and further reduced in a Ca^{2+}-free environment that was established by addition of EGTA (Figure 5-7 B). This observation confirmed a previous study by Gönczi *et al.* [60] reporting that HaCaT cells respond to hypotonic stimulation by increase of intracellular Ca^{2+}, which critically depends on extracellular Ca^{2+}.

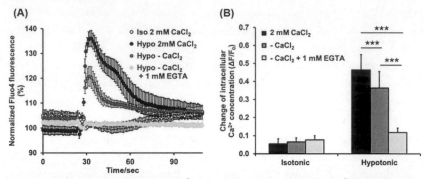

Figure 5-7: Effect of the extracellular Ca^{2+} concentration on intracellular Ca^{2+} concentration

HaCaT cells were loaded with Ca^{2+}-indicator Fluo4-AM prior to measurement and stimulated with isotonic (329 mOsm) and hypotonic (150 mOsm) buffers containing either 2 mM $CaCl_2$, no $CaCl_2$ (- $CaCl_2$) or no $CaCl_2$ but 1 mM EGTA. (A) Fluo4 fluorescence was normalized to fluorescence at the time point of stimulation and plotted over time showing increase of intracellular Fluo4 fluorescence upon hypotonic stimulation. One representative experiment is presented depicting the mean with SD from four technical replicates. (B) Maximal increase in Fluo4 fluorescence ($\Delta F = F_{max} - F_0$) was normalized to baseline fluorescence (F_0) and used as a measure of change of intracellular Ca^{2+} concentration ($\Delta F/ F_0$). Increase of intracellular Ca^{2+} concentration depends on presence of extracellular Ca^{2+}. Bars represent the mean with SD of at least 58 measurements from three independent experiments.

5.3.2   VRAC activity is modulated by Ca^{2+}

Since the intracellular Ca^{2+} concentration increased upon hypotonic stress in dependence of extracellular Ca^{2+} concentration, it was investigated how Ca^{2+} affects VRAC activity. Therefore, hsYFP quenching was measured to determine VRAC activity upon hypotonic stimulation with buffers of different Ca^{2+} concentrations.

Strongest quenching of hsYFP fluorescence (10 %) was monitored upon hypotonic stimulation in the presence of 2 mM $CaCl_2$. Hypotonic stimulation with buffer without addition of $CaCl_2$ resulted in reduced hsYFP fluorescence quenching (7 %) that was further reduced (4 %) when residual extracellular Ca^{2+} traces were chelated by EGTA. Isotonic stimulation did not change hsYFP fluorescence (Figure 5-8 A). These data show that with decreasing extracellular Ca^{2+} concentration also quenching of hsYFP fluorescence decreased.

Accordingly, highest VRAC activity was calculated in the presence of 2 mM $CaCl_2$ (Figure 5-8 B) when also the strongest elevation of intracellular Ca^{2+} concentration was detected (Figure 5-7 B). Stimulation without adding $CaCl_2$ reduced VRAC activity as well as increase in intracellular Ca^{2+} (Figure 5-7 B). Importantly, VRAC activity was still detected in the absence of extracellular Ca^{2+} (- $CaCl_2$ + 1 mM EGTA in Figure 5-8 B) and without elevation of intracellular Ca²⁺ (Figure 5-7 B). In conclusion, VRAC activity does not require Ca²⁺, but increase in extra- and intracellular Ca²⁺ concentration increases VRAC activity.

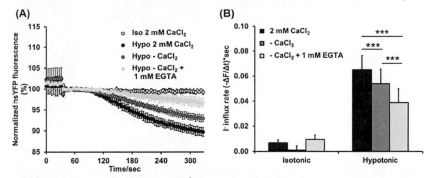

Figure 5-8: Effect of the extracellular Ca²⁺ concentration on VRAC activity

HaCaT-*hsYFP* cells were stimulated with isotonic (329 mOsm) and hypotonic (229 mOsm) buffers containing either 2 mM CaCl₂, no CaCl₂ (- CaCl₂) or no CaCl₂ but 1 mM EGTA. (A) HsYFP fluorescence was normalized to fluorescence at the time point of stimulation and plotted over time. HsYFP fluorescence decreased upon hypotonic stimulation containing 2 mM CaCl₂, but less when CaCl₂ was reduced or omitted. One representative experiment is shown depicting the mean with SD from five technical replicates. (B) I influx rate (-ΔF/Δt) was calculated as a measure of VRAC activity. Highest VRAC activity was calculated in presence of 2 mM CaCl₂ and lowest in Ca²⁺-free condition (- CaCl₂ + 1 mM EGTA). Bars represent the mean with SD of at least 60 measurements from three independent experiments.

5.3.3   RVD is accelerated by extracellular Ca²⁺

As extracellular Ca²⁺ modulates VRAC activity and VRACs are activated during RVD, it was investigated whether Ca²⁺ influences RVD. RVD of HaCaT cells was monitored by concentration-sensitive fluorophore calcein as described above. Calcein fluorescence increased upon hypotonic stimulation and decreased the fastest and below baseline level when 2 mM extracellular CaCl₂ was present. Fluorescence decrease was slower but still reaching baseline when extracellular Ca²⁺ was omitted (- CaCl₂). In Ca²⁺-free buffer (- CaCl₂ + 1 mM EGTA) calcein fluorescence decreased even slower not reaching baseline level (Figure 5-9 A). Interestingly, cell swelling, monitored by fluorescence increase, was independent from extracellular Ca²⁺ concentration (Figure 5-9 A).

The speed of RVD (-ΔF/Δt) was calculated from calcein fluorescence decrease over time. Importantly, cells shrank in a Ca²⁺-free environment (- CaCl₂ + 1 mM EGTA) indicating that Ca²⁺ is not essential for RVD. However, RVD occurred faster in the presence of extracellular Ca²⁺ traces (- CaCl₂) and even faster in the presence of 2 mM CaCl₂ (Figure 5-9 B). In agreement with VRAC activity, RVD is Ca²⁺-independent but accelerated by Ca²⁺-dependent processes.

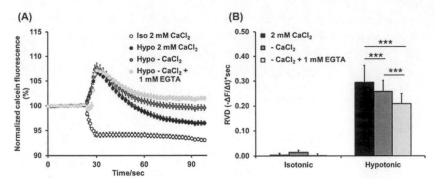

Figure 5-9: Effect of the extracellular Ca²⁺ concentration on regulatory volume decrease

HaCaT cells were loaded with concentration-sensitive calcein-AM prior to measurement and stimulated with isotonic (329 mOsm) and hypotonic (150 mOsm) buffers containing either 2 mM CaCl₂, no CaCl₂ (- CaCl₂) or no CaCl₂ but 1 mM EGTA. (A) Calcein fluorescence was normalized to fluorescence at the time point of stimulation and plotted over time. Increase of calcein fluorescence is independent from extracellular Ca²⁺ concentration, but fluorescence decrease is modulated by extracellular Ca²⁺. One representative experiment is shown depicting the mean with SD from five technical replicates. (B) Fluorescence decrease over time (-ΔF/Δt) was used as a measure of speed of RVD. RVD occurred fastest in the presence of 2 mM CaCl₂, but was decelerated by reduction of extracellular CaCl₂. Bars represent the mean with SD of 60 measurements from three independent experiments.

To directly compare the relationship of intracellular Ca²⁺ concentration, VRAC activity and speed of RVD, all three parameters were normalized as described in the methods section (7.3.3, 7.3.1 and 7.3.2). When CaCl₂ was not added to hypotonic buffer, increase of intracellular Ca²⁺ concentration was reduced to 77 %. Interestingly, increase of intracellular Ca²⁺ concentration was abolished upon Ca²⁺-free stimulation (- CaCl₂ + 1 mM EGTA) (Figure 5-10 A). These data suggest that Ca²⁺ originates from intracellular stores whose activation however depends on traces of extracellular Ca²⁺. VRAC activity was reduced to 80 % or 53 % by reduction (- CaCl₂) or abolishment (- CaCl₂ + 1 mM EGTA) of extracellular Ca²⁺, respectively (Figure 5-10 B). This observation can be correlated with the change in intracellular Ca²⁺ concentration that also decreased with decreasing extracellular Ca²⁺. Omitting CaCl₂ from hypotonic buffer resulted in reduction of RVD to 89 % while abolishment of Ca²⁺ (- CaCl₂ + 1 mM EGTA) further reduced RVD to 71 % (Figure 5-10 C). This reduction could be a direct result of the reduction of VRAC activity. In summary, reduction of extracellular Ca²⁺ reduced the increase in intracellular Ca²⁺, which correlates with reduction of VRAC activity as well as RVD.

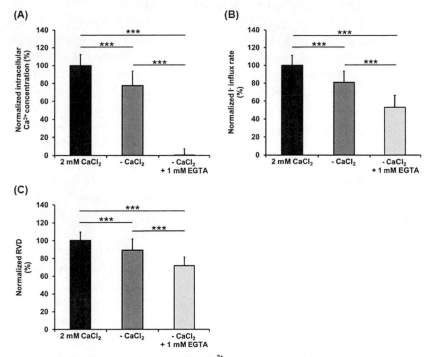

Figure 5-10: Dependence of intracellular Ca²⁺ concentration, I⁻ influx rate and RVD on extracellular Ca²⁺concentration

Change in intracellular Ca²⁺ concentration (A), I⁻ influx rate (B) and RVD (C) upon hypotonic stimulation (150 mOsm (A and C) or 229 mOsm (B)) with different CaCl₂ concentrations were calculated and related to stimulation with buffer containing 2 mM CaCl₂. (A) Fluo4 loaded HaCaT cells were used to measure changes of intracellular Ca²⁺ concentration. (B) I⁻ influx rate was calculated from hsYFP quenching of HaCaT-*hsYFP*. (C) Calcein loaded HaCaT cells were used to measure RVD. Change in intracellular Ca²⁺ concentration, I⁻ influx rate and RVD depend on presence of extracellular CaCl₂.

5.4   LRRC8A is essential for VRAC activity and involved in RVD of human keratinocytes

After expression of *LRRC8A* in native human epidermis as well as in cultivated keratinocytes was demonstrated and after having shown that VRACs are activated during hypotonic stress response, it was hypothesized that LRRC8A is mediating VRAC activity and subsequent RVD also in keratinocytes. To answer this question human keratinocytes devoid of *LRRC8A* were generated using the CRISPR-Cas9 technology. Genome-edited *LRRC8A* knock-out keratinocytes were then used to investigate VRAC activity and RVD.

5.4.1 Generation of HaCaT keratinocytes devoid of *LRRC8A*

To generate HaCaT keratinocytes devoid of *LRRC8A* a variety of single-guide RNAs (sgRNA) were designed and validated by tracking of indels by decomposition (TIDE) analysis for their efficiency to generate Cas9-induced genomic mutations. Mutations occur by error-prone non-homologous end joining of double-strand breaks that are induced by the Cas9 nuclease, which is guided by sgRNAs to their corresponding specific sites in the genome. The two most efficient sgRNAs were then chosen that target the coding sequence (CDS) of *LRRC8A* downstream and upstream of the first transmembrane domain (Figure 5-11 A). By targeting the start of the CDS of *LRRC8A* it can be assured that CRISPR-Cas9 mediated genome editing results in a dysfunctional protein. When the Cas9 nuclease cleaves at both targeted sites at the same time a genomic region of 295 bp is deleted and a frameshift mutation is caused. Vectors encoding both guide RNAs and the Cas9 nuclease were delivered into HaCaT cells by adenoviral transduction and limiting dilution was used to generate monoclonal cell lines devoid of LRRC8A.

In order to characterize the generated cell lines, genomic DNA of these cells was subjected to target site-specific PCR. Amplification of the *LRRC8A* target region of cells without editing or single nucleotide insertions/deletions resulted in a PCR product of approximately 660 bp while an approximately 360 bp long PCR product was detected for cells with the predicted genomic deletion in the *LRRC8A* locus (Figure 5-11 B). Subsequent Sanger sequencing of the PCR products detected the genomic alteration and one cell line devoid of *LRRC8A* was chosen (HaCaT-*LRRC8A*$^{-/-}$). As control, a clonal wild-type cell line was chosen, which was treated with the same adenovirus, however, without alteration in the *LRRC8A* locus. Sequencing of the PCR product from HaCaT-*LRRC8A*$^{-/-}$ cells confirmed the predicted cleavage after cytidine 27 and a deletion of 295 bp or 305 bp on the two alleles (Figure 5-11 D and E). *In silico* translation predicted that this deletion results in a frameshift mutation after phenylalanine 10 and a premature stop codon (Figure 5-11 E) so that no functional protein can be formed. Indeed, Western blot analysis confirmed the absence of LRRC8A protein in HaCaT-*LRRC8A*$^{-/-}$ cells (Figure 5-11 C).

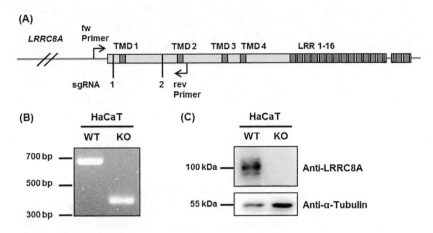

(A)

LRRC8A — fw Primer — TMD 1 — TMD 2 — TMD 3 TMD 4 — LRR 1-16

sgRNA 1 — 2 — rev Primer

(B) HaCaT WT KO

700 bp
500 bp
300 bp

(C) HaCaT WT KO

100 kDa — Anti-LRRC8A
55 kDa — Anti-α-Tubulin

(D)

Cells	Allele	Genetic modification	Protein modification
HaCaT-LRRC8A⁻/⁻	1	Δ295 nt (c27 – c321)	F10T and additional frameshift, stop after aa62: aa1-61
	2	Δ305 nt (c27 – g331)	F10C and additional frameshift, stop after aa12: aa1-12

(E)

```
                                        PAM       sgRNA1
WT                 atgattccggtgacagagctccgctactttgcggacacgcagccagcata
LRRC8A KO Allele 1 atgattccggtgacagagctccgcta------------------------
LRRC8A KO Allele 2 atgattccggtgacagagctccgcta------------------------

WT                 ccggatcctgaagccgtggtgggatgtgttcacagactacatctctatcg
LRRC8A KO Allele 1 --------------------------------------------------
LRRC8A KO Allele 2 --------------------------------------------------

WT                 tcatgctgatgattgccgtcttcggggggacgctgcaggtcacccaagac
LRRC8A KO Allele 1 --------------------------------------------------
LRRC8A KO Allele 2 --------------------------------------------------

WT                 aagatgatctgcctgccttgtaagtgggtcaccaaggactcctgcaatga
LRRC8A KO Allele 1 --------------------------------------------------
LRRC8A KO Allele 2 --------------------------------------------------

WT                 ttcgttccggggctgggcagcccctggcccggagcccacctaccccaact
LRRC8A KO Allele 1 --------------------------------------------------
LRRC8A KO Allele 2 --------------------------------------------------

WT                 ccaccattctgccgacccctgacacgggccccacaggcatcaagtatgac
LRRC8A KO Allele 1 --------------------------------------------------
LRRC8A KO Allele 2 --------------------------------------------------

                            sgRNA 2            PAM
WT                 ctggaccggcaccagtacaactacgtggacgctgtgtgctatgagaaccg
LRRC8A KO Allele 1 --------------------tacgtggacgctgtgtgctatgagaaccg
LRRC8A KO Allele 2 ----------------------------ctgtgtgctatgagaaccg
```

Figure 5-11: Generation of HaCaT-LRRC8A⁻/⁻ cells

----- continued on the next page -----

(A) Illustration of the *LRRC8A* gene depicting the location of the four transmembrane domains (TMD) and 16 leucine-rich repeat domains (LRR). Position of the two single-guide RNAs (sgRNAs) that guide the Cas9 nuclease and binding sites for target site-specific PCR primers (fw and rev primer) are shown. (B) Genomic DNA from HaCaT-WT (WT) and HaCaT-*LRRC8A*^-/- (KO) was subjected to target site-specific PCR. Amplification products of approximately 660 bp (WT) and 360 bp (KO) were visualized. (C) Whole cell protein lysates from HaCaT-WT (WT) and HaCaT-*LRRC8A*^-/- (KO) were used for Western blot analysis to detect LRRC8A protein by a LRRC8A-specific antibody. α-tubulin was used as loading control. (D) Sequencing of the amplification products showed different genomic modification at both *LRRC8A* alleles that result in different protein modifications. (E) Alignment of the PCR products from HaCaT-WT and both alleles of HaCaT-*LRRC8A*^-/- cells illustrates the deletion of 295 bp on allele 1 and 305 bp on allele 2. SgRNAs and PAMs are depicted.

5.4.2 VRAC activity is abolished in HaCaT-*LRRC8A*^-/- cells

To address whether LRRC8A mediates VRAC activity of HaCaT keratinocytes, VRAC activity was measured by hypotonicity-induced quenching of hsYFP fluorescence. Therefore, HaCaT-WT and HaCaT-*LRRC8A*^-/- cells were transduced with an adenovirus to deliver the gene expression cassette for the iodide sensor hsYFP prior to measurement.

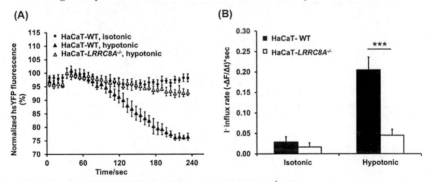

Figure 5-12: Reduction of VRAC activity in HaCaT-*LRRC8A*^-/- cells

HaCaT-WT and HaCaT-*LRRC8A*^-/- cells were transduced to express *hsYFP* as iodide sensor prior to isotonic or hypotonic stimulation. (A) HsYFP fluorescence was normalized to the time point of stimulation and plotted over time. Hypotonic stimulation of HaCaT-WT cells resulted in 25 % quenching of hsYFP fluorescence, which was reduced to level of isotonic control in HaCaT-*LRRC8A*^-/- cells. One representative experiment is shown depicting the mean with SD from five technical replicates. (B) hsYFP fluorescence decrease over time ($-\Delta F/\Delta t$) was calculated as measure of I^- influx rate showing reduced I^- influx rate in HaCaT-*LRRC8A*^-/- cells. Bars represent the mean with SD of 24 (isotonic) or 71 (hypotonic) measurements from three independent experiments.

Hypotonic stimulation of HaCaT-WT cells led to fast I^- influx and subsequent quenching of hsYFP fluorescence by 25 %. Strikingly, in HaCaT cells lacking LRRC8A decrease of hsYFP fluorescence was only 5 % and not very different from isotonic control (3 %) (Figure 5-12 A). Quantification of I^- influx rate as measure of VRAC activity showed that VRAC activity in the absence of LRRC8A was reduced by approximately 75 % and only two-times higher than upon isotonic control stimulation (Figure 5-12 B). These data clearly show that LRRC8A is an essential component of the volume-regulated anion channel in HaCaT keratinocytes.

5.4.3   RVD is reduced in HaCaT-*LRRC8A*$^{-/-}$ cells

Although it was shown that VRACs in general and LRRC8A in particular mediate RVD in different cell types, the degree of contribution is variable [88, 89] and probably cell type-dependent. Since LRRC8A is essential for VRAC activity in HaCaT keratinocytes, it was investigated whether LRRC8A also contributes to RVD of HaCaT cells.

Cell volume changes of HaCaT-WT and HaCaT-*LRRC8A*$^{-/-}$ cells were monitored using calcein-AM. Hypotonicity-induced cell volume increase and subsequent decrease were detected in HaCaT-WT cells by 8 % increase of calcein fluorescence and subsequent decrease to baseline fluorescence. Hypotonic stimulation of HaCaT-*LRRC8A*$^{-/-}$ cells resulted in 5 % fluorescence increase. Importantly, absence of LRRC8A resulted in only 2 % decrease of calcein fluorescence and is therefore slower than in HaCaT-WT cells and incomplete

(Figure 5-13 A) indicating impaired shrinkage of HaCaT cells. The speed of RVD was calculated from the decrease of calcein fluorescence over time. Importantly, RVD of HaCaT cells devoid of LRRC8A was decelerated by half (Figure 5-13 B). Taken together, in the absence of LRRC8A RVD occurred slower and incomplete suggesting that LRRC8A is contributing to RVD in HaCaT keratinocytes.

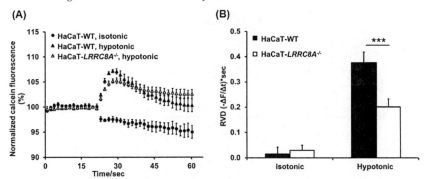

Figure 5-13: Reduction of RVD in HaCaT-LRRC8A$^{-/-}$ cells

HaCaT-WT and HaCaT-*LRRC8A*$^{-/-}$ cells were loaded with concentration-sensitive calcein-AM prior to isotonic or hypotonic stimulation. (A) Calcein fluorescence was normalized to the time point of stimulation and plotted over time. Slower and incomplete decrease of calcein fluorescence was monitored for HaCaT-*LRRC8A*$^{-/-}$ cells. One representative experiment is shown depicting the mean with SD from five technical replicates. (B) The decrease of calcein fluorescence over time was calculated as a measure of RVD showing a 50 % reduction of RVD in HaCaT-*LRRC8A*$^{-/-}$ cells. Bars represent the mean with SD of 36 (isotonic) or 89 (hypotonic) measurements from three independent experiments.

5.4.4 Generation of *LRRC8A* knock-out in primary keratinocytes

The importance of LRRC8A as essential VRAC component that mediates RVD in the HaCaT keratinocyte cell line should be corroborated in primary keratinocytes. Due to the short lifespan of primary cells of only a few population doublings, it is not possible to raise single cell clones from primary cells. To circumvent this problem, functional analyses were performed in a mixed cell population containing different genomic alterations.

To generate primary keratinocyte populations devoid of *LRRC8A*, normal human epidermal keratinocytes (NHEK) were transduced with adenoviruses, which delivered the gene expression cassettes for the Cas9 nuclease and two guide RNAs targeting *LRRC8A* as described for HaCaT cells. As control, NHEK cells were transduced only with Cas9 nuclease without sgRNA. Three independent cell populations were generated and genomic DNA and whole cell lysates were prepared. Genomic DNA was subjected to target site-specific PCR and sequencing of the PCR products showed that approximately 55 % of cells carried a genomic alteration; either the predicted genomic deletion of 295 nt or insertions or deletions of single nucleotides at either one of the Cas9-targeted positions (Figure 5-14 A). Whole cell lysates were subjected to Western blot analysis (Figure 5-14 B) and densitometric quantification of all three Western blots demonstrated on average a residual LRRC8A protein content of 21 % compared to NHEK-WT.

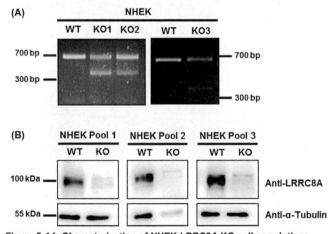

Figure 5-14: Characterization of NHEK-LRRC8A KO cell populations

NHEK cells were transduced with adenoviruses to deliver the gene expression cassettes either for only Cas9 nuclease (WT) or additionally two sgRNAs directed against *LRRC8A* to generate NHEK-*LRRC8A* KO pools (KO). (A) Genomic DNA from three independently generated NHEK-WT (WT) and NEHK-*LRRC8A* KO pools (KO) was subjected to target site-specific PCR resulting in amplification products of approximately 660 bp for WT or 660 bp and 360 bp for KO pools. (B) Whole cell lysates of the three NHEK cell pools (Pool 1-3) were used for Western blot analysis to detect LRRC8A protein. α-tubulin was used as loading control.

5.4.5 Reduced VRAC activity in NHEK-*LRRC8A* KO cells

In order to determine VRAC activity of primary keratinocytes with reduced LRRC8A level, hypotonicity-induced quenching of the iodide sensor hsYFP was monitored. NHEK-WT and NHEK-*LRRC8A* KO cell populations were transduced with an adenovirus to deliver the gene expression cassette for *hsYFP* prior to measurement. For NHEK-WT 20 % quenching of hsYFP fluorescence within 190 sec was detected upon hypotonic stimulation, but none upon isotonic stimulation. Importantly, speed and extend of hsYFP quenching upon hypotonic stimulation was diminished to a maximum of 10 % hsYFP quenching within 95 sec in primary keratinocytes with reduced LRRC8A level (Figure 5-15 A).

I⁻ influx rate was calculated as measure of VRAC activity and was readily detected in NHEK-WT cells. With a residual LRRC8A protein level of around 20 % in NHEK-*LRRC8A* KO cells, VRAC activity was reduced by two third after KO of LRRC8A (Figure 5-15 B). These results strengthen the finding that LRRC8A is an essential component of VRACs in both, HaCaT and primary keratinocytes.

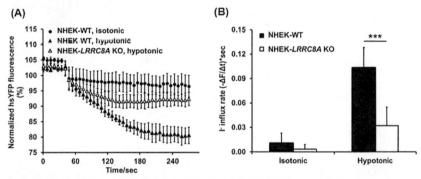

Figure 5-15: Reduction of VRAC activity in NHEK-LRRC8A KO cell populations

NHEK-WT and NHEK-*LRRC8A* KO pools were transduced to express hsYFP. (A) hsYFP fluorescence was normalized to the time point of stimulation and plotted over time. HsYFP quenching was reduced in NHEK-*LRRC8A* KO cells compared to NHEK-WT cells. One representative experiment is shown depicting the mean with SD from five technical replicates. (B) I⁻ influx rate was calculated as measure of VRAC activity showing a reduction to one third in NHEK-*LRRC8A* KO cells. Bars represent the mean with SD of 11 (isotonic) and 28 (hypotonic) measurements from three independent experiments.

5.4.6 Reduced RVD in NHEK-*LRRC8A* KO cells

Since the important function of LRRC8A as VRAC in primary keratinocytes was shown, the impact on RVD was investigated. Therefore, NHEK-WT and NHEK-*LRRC8A* KO cells were loaded with calcein-AM and fluorescence changes upon isotonic and hypotonic stimulation were measured.

Hypotonic stimulation of NHEK-WT resulted in increase of calcein fluorescence, which, after reaching a maximum at 108 %, decreased to baseline fluorescence indicating cell swelling and subsequent cell volume decrease. A comparable increase of calcein fluorescence was monitored in NHEK-*LRRC8A* KO cells. However, calcein fluorescence decreased less and stayed elevated over baseline level (104 %) (Figure 5-16 A) indicating slower and incomplete cell volume decrease. The speed of RVD was quantified by calcein fluorescence decrease over time. NHEK-WT cells responded to hypotonic stimulation with fast RVD, which was decelerated by one third in primary keratinocytes with reduced LRRC8A level (Figure 5-16 B). These findings confirm the important role of LRRC8A in regulatory volume decrease of cultured keratinocytes upon hypotonic stress.

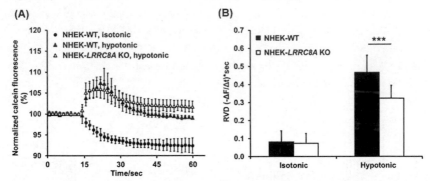

Figure 5-16: Reduction of RVD in NHEK-LRRC8A KO cells

NHEK-WT and NHEK-*LRRC8A* KO cells were loaded with calcein-AM prior to stimulation. (A) Calcein fluorescence was normalized to the time point of stimulation and plotted over time showing slower and incomplete cell volume decrease in NHEK-*LRRC8A* KO cells. One representative experiment is depicted showing the mean with SD from five technical replicates. (B) RVD was calculated from calcein fluorescence decrease after maximal fluorescence over time. RVD was reduced by one third in NHEK-*LRRC8A* KO cells. Bars represent the mean with SD of 32 (isotonic) and 81 (hypotonic) measurements from three independent experiments.

Iodide influx rate and speed of RVD were normalized to better compare the impact of LRRC8A on VRAC activity and RVD. Strikingly, VRAC activity of HaCaT-*LRRC8A*[-/-] cells was reduced by 90 % (Figure 5-17 A) and 50 % reduction in regulatory volume decrease was achieved (Figure 5-17 B). Furthermore, primary NHEK cells with 20 % residual LRRC8A protein showed a reduction of 70 % in VRAC activity (Figure 5-17 A) as well as 35 % decrease in RVD (Figure 5-17 B). Taken together, it was revealed that LRRC8A is an essential component of VRACs also in HaCaT and primary keratinocytes and LRRC8A contributes to regulatory volume decrease upon hypotonic stress.

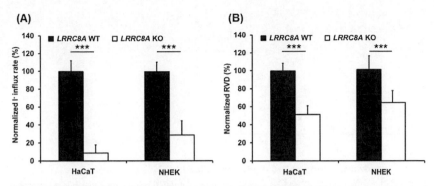

Figure 5-17: Reduction of iodide influx rate and RVD of keratinocytes by knock-out of LRRC8A

(A) Iodide influx rate as measure of VRAC activity of KO cells as related to iodide influx rate of WT cells and presented in percent. Reduction by 90 % (HaCaT-*LRRC8A*^{-/-}) and 70 % (NHEK-*LRRC8A* KO cells with residual protein level of 20 %) shows that LRRC8A is essential for VRAC activity. (B) RVD of KO cells was related to RVD of WT cells and presented in percent. LRRC8A contributes to regulatory volume decrease in keratinocytes as shown by 50 % and 35 % reduction of RVD in HaCaT and NEHK cells.

5.5   LRRC8A is involved in proliferation of HaCaT keratinocytes

In native epidermis, only basal keratinocytes proliferate in order to ensure a constant epidermal renewal. Interestingly, LRRC8A seems to be expressed only in basal keratinocytes (Figure 5-1) thus raising the question whether LRRC8A is involved in keratinocyte proliferation. This consideration is further supported by the fact, that cell volume changes are crucial for cell proliferation [54] and LRRC8A is mediating cell volume decrease upon hypotonic stress. To investigate the role of LRRC8A in keratinocyte proliferation, proliferation of HaCaT-WT and HaCaT-*LRRC8A*^{-/-} cells was measured by different approaches.

First, proliferation was assessed by acquiring a growth curve. A defined number of cells was seeded and cell number was determined by automated cell counting every 24 h for five days. Cell number was plotted over time to generate a growth curve, which showed no difference in cell number of HaCaT-WT and HaCaT-*LRRC8A*^{-/-} cells for the first 72 h. However, after 96 h of culture a 30 % lower cell number was determined for HaCaT-*LRRC8A*^{-/-} cells compared to HaCaT-WT. That difference increased to 60 % less HaCaT-*LRRC8A*^{-/-} than HaCaT-WT cells after 120 h (Figure 5-18).

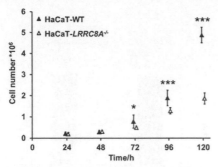

Figure 5-18: Growth curves of HaCaT-WT and HaCaT-LRRC8A$^{-/-}$ cells

HaCaT-WT and HaCaT-*LRRC8A*$^{-/-}$ cells were seeded in a defined number of cells and cell numbers were determined by automated cell counting every 24 h for five days. Cell number of HaCaT-WT cells increased stronger compared to HaCaT-*LRRC8A*$^{-/-}$ cells indicating a proliferation deficit in the absence of LRRC8A. The graph depicts the mean with SD from 18 measurements acquired in three independent experiments. Asterisks depict the statistical significance between number of HaCaT-WT and HaCaT-*LRRC8A*$^{-/-}$ cells.

From the exponential increase in cell number the population doubling time was calculated. HaCaT-WT cells required 18.9 ± 0.3 h for one population doubling, which was increased for HaCaT-*LRRC8A*$^{-/-}$ cells to 24.3 ± 1.0 h. The lower cell number and prolonged doubling time of HaCaT-*LRRC8A*$^{-/-}$ compared to HaCaT-WT cells led to the hypothesis that LRRC8A is involved in HaCaT cell proliferation.

To further corroborate the role of LRRC8A in proliferation two additional methods were used. One method is based on the conversion of the water soluble tetrazolium salt WST-1 to formazan by dehydrogenases of living cells thereby depicting the metabolic activity of the cells. The absorbance of formazan is measured and correlates with the number of vital cells [123]. The second method is an immunological detection of the synthetic thymidine analogue bromodeoxyuridine (BrdU) [124]. BrdU is incorporated into the DNA during DNA replication and specifically bound by a horseradish peroxidase-coupled Anti-BrdU antibody. Finally, BrdU incorporation is detected by absorption of the oxidized chromogenic substrate tetramethylbenzidine. Metabolic activity and DNA synthesis of HaCaT-WT and HaCaT-*LRRC8A*$^{-/-}$ cells was measured 120 h after cell seeding, when the growth curve showed the greatest difference in cell number (Figure 5-18). Absorption of formazan and tetramethylbenzidine was reduced by 40 % and 29 %, respectively (Figure 5-19) indicating a reduction of metabolic activity and DNA synthesis of HaCaT-*LRRC8A*$^{-/-}$ cells. In summary, it was shown that HaCaT keratinocytes devoid of *LRRC8A* show a reduced growth rate, metabolic activity and DNA synthesis indicating that LRRC8A is involved in proliferation of HaCaT keratinocytes.

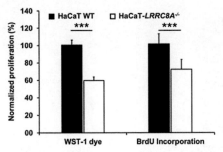

Figure 5-19: Reduced proliferation of HaCaT-LRRC8A$^{-/-}$ cells

Metabolic activity and DNA synthesis of HaCaT-WT and HaCaT-*LRRC8A*$^{-/-}$ cells were assessed by absorption of formazan or oxidized tetramethylbenzidine after conversion of WST-1 or BrdU incorporation, respectively. Absorption was measured and related to values obtained from HaCaT-WT cells. In the absence of LRRC8A metabolic activity as well as DNA synthesis was reduced. Bars represent the mean with SD of 53 (WST-1) or 34 (BrdU Incorporation) measurements from three or two independent experiments, respectively.

5.6   LRRC8A modifies differentiation of HaCaT keratinocytes

Keratinocytes that leave the basal layer switch from proliferation to differentiation, which is associated with an increase in cell size. In this study, it was shown that LRRC8A is preferentially expressed in basal rather than differentiating keratinocytes in native epidermis. Furthermore, HaCaT keratinocytes devoid of LRRC8A show impaired cell volume regulation and a reduced proliferative capacity. These aspects hint to an involvement of LRRC8A in keratinocyte differentiation, which was subsequently investigated.

5.6.1   Differentiation of HaCaT keratinocytes by post-confluent growth

Differentiation of HaCaT keratinocytes can be induced by post-confluent growth [15]. HaCaT cells were seeded in different cell densities ranging from 0.1*10⁶ to 1*10⁶ cells per 6-well cavity and grown for two days. At low cell densities HaCaT cells grew only in a few small colonies while a post-confluent cell layer could be achieved for higher cell densities (Figure 5-20). At all cell densities HaCaT cells appeared small and edged, and grew in tight colonies. Morphology of HaCaT-WT and HaCaT-*LRRC8A*$^{-/-}$ cells was comparable.

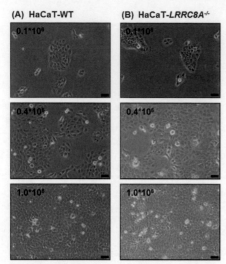

(A) HaCaT-WT   **(B) HaCaT-*LRRC8A*-/-**

Figure 5-20: Microscopic images of HaCaT-WT and HaCaT-LRRC8A⁻/⁻ cells during post-confluent growth

HaCaT-WT (A) and HaCaT-*LRRC8A*-/- cells (B) were seeded at indicated cell numbers and grown for two days to induce differentiation prior to microscopic analysis. Confluency increases and cell size decreases with increasing cell number. Bars represent 50 μm.

To evaluate if LRRC8A influences differentiation of HaCaT keratinocytes, gene expression of epidermal differentiation markers was analyzed by RNA sequencing. Total mRNA was isolated from HaCaT-WT and HaCaT-*LRRC8A*-/- cells grown at lowest (0.1*10⁶ cells/cavity) as well as highest (1.0*10⁶ cells/cavity) cell density to represent early and late differentiation state. The mRNA abundance of differentiation-specific genes in HaCaT-WT and HaCaT-*LRRC8A*-/- cells was compared. *Keratin 5, keratin 14, keratin 1, keratin 10, involucrin, filaggrin* and *transglutaminase 1* are respective markers that are expressed in different epidermal layers.

Keratin 5 (KRT5) and *keratin 14 (KRT14)* are endogenously expressed only in basal keratinocytes [125, 126]. In HaCaT-WT and HaCaT-*LRRC8A*-/- cells however, expression of both genes increased from lowest to highest cell density and expression of *keratin 5* in HaCaT-*LRRC8A*-/- cells was increased compared to HaCaT-WT. (Figure 5-21 A). The first genes that are specific for differentiating keratinocytes are *keratin 1 (KRT1)* and *keratin 10 (KRT10)* [127]. Upon post-confluent growth of HaCaT-WT and HaCaT-*LRRC8A*-/- cells *keratin 1* and *keratin 10* increased from early to late differentiation state (Figure 5-21 B). *Filaggrin (FLG), involucrin (IVL)* and *transglutaminase 1 (TGM1)* are expressed in further differentiated keratinocytes of the epidermis. mRNA abundance of *filaggrin* decreased from early to late differentiation state in HaCaT-WT cells, but stayed constant in HaCaT-*LRRC8A*-/- cells. As seen for *keratin 1* and *keratin 10*, also *involucrin* and *transglutaminase 1* increased from early to late differentiation state of HaCaT-WT and HaCaT-*LRRC8A⁻/⁻* cells (Figure 5-21 B and C). Interestingly, in HaCaT-*LRRC8A⁻/⁻* cells *keratin 1, keratin 10, involucrin* and *transglutaminase 1* were higher expressed compared to HaCaT-WT cells indicating a higher degree of differentiation in the absence of LRRC8A.

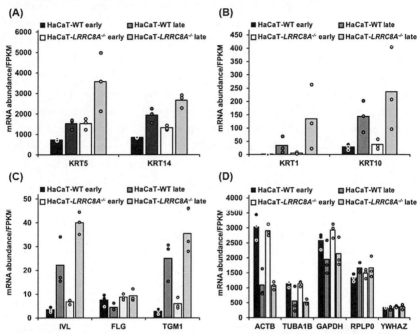

Figure 5-21: RNA sequencing of HaCaT-WT and HaCaT-LRRC8A⁻/⁻ cells upon post-confluent growth

HaCaT-WT and HaCaT-*LRRC8A⁻/⁻* cells were differentiated by post-confluent growth and total mRNA from early and late differentiation state was isolated and subjected to RNA sequencing. Basal markers *keratin 5* (*KRT5*) and *keratin 14* (*KRT 14*), differentiation markers *keratin 1* (*KRT1*), *keratin 10* (*KRT10*), *involucrin* (*IVL*), *transglutaminase 1* (*TGM1*) and *filaggrin* (*FLG*) as well as house-keeping genes *β-actin* (*ACTB*), *glyceraldehyde-3-phosphate dehydrogenase* (*GAPDH*), *α-tubulin* (*TUBA1B*), *ribosomal protein lateral stalk subunit P0* (*RPLP0*) and *tyrosine 3-monooxygenase/tryptophan 5-monooxygenase activation protein zeta* (*YWHAZ*) were analyzed. Transcript abundances were displayed as the mean (bar) of the individual values (dots) from three independent RNA preparations in fragments per kilobase of exon per million fragments mapped (FPKM).

Additionally, expression of typical house-keeping genes *β-actin* (*ACTB*), *glyceraldehyde-3-phosphate dehydrogenase* (*GAPDH*) and *α-tubulin* (*TUBA1B*) decreased during HaCaT differentiation. Lanzafame *et al.* already described this phenomenon and proposed *ribosomal protein lateral stalk subunit P0* (*RPLP0*) or *tyrosine 3-monooxygenase/tryptophan 5-monooxygenase activation protein zeta* (*YWHAZ*) as most stable genes during keratinocyte differentiation [128]. RNA sequencing indeed showed stable expression of these two genes in early and late differentiation state of HaCaT-WT and HaCaT-*LRRC8A^-/-* cells (Figure 5-21 D). Therefore, RPLP0 and YWHAZ were used as a loading control in Western blot experiments.

To complement gene expression analysis, involucrin and keratin 10 as markers for differentiation were also analyzed on protein level by Western blotting. Whole cell protein lysates were isolated from HaCaT cells seeded in cell densities from $0.1*10^6 - 1.0*10^6$ cells/cavity. To analyze the dynamics of differentiation, cell densities between $0.1*10^6$ (early differentiation) and $1.0*10^6$ (late differentiation) were included. Western blot showed that with increasing cell density involucrin and keratin 10 increased (Figure 5-22) confirming differentiation of HaCaT keratinocytes by post-confluent growth. Interestingly, in HaCaT-*LRRC8A^-/-* cells, both differentiation markers already appeared at a lower cell density as well as with stronger intensity at highest cell density (Figure 5-22 and Supplement Figure S2). This points towards an earlier onset of differentiation in HaCaT keratinocytes devoid of LRRC8A. Western blot analyses support the putative involvement of LRRC8A in differentiation of HaCaT keratinocytes suggested by RNA sequencing data.

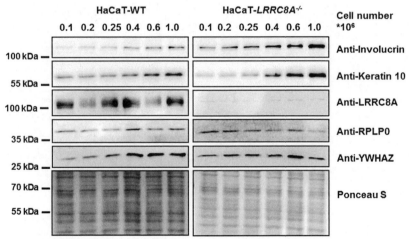

Figure 5-22: Differentiation of HaCaT-WT and HaCaT-LRRC8A^-/- cells upon post-confluent growth

HaCaT-WT and HaCaT-*LRRC8A^-/-* cells were seeded in increasing cell densities from $0.1*10^6 - 1.0*10^6$ cells/cavity. Whole cell protein lysates were subjected to Western blot analysis. The differentiation markers involucrin and keratin 10 increased with increasing cell density indicating differentiation of HaCaT cells. Earlier appearance and stronger intensity of involucrin and keratin 10 in HaCaT-*LRRC8A^-/-* cells point to involvement of LRRC8A in HaCaT keratinocyte differentiation. Absence of LRRC8A in HaCaT-*LRRC8A^-/-* cells was confirmed. Western blot analysis of RPLP0 and YWHAZ were used as loading control as well as membrane staining using Ponceau S.

5.6.2 Differentiation of HaCaT keratinocytes by extracellular Ca^{2+} concentration

Differentiation state of HaCaT keratinocytes cannot only be influenced by cell density, but also by the Ca^{2+} concentration in the culture medium. HaCaT cells have to be cultured for three weeks in medium containing only 0.03 mM $CaCl_2$ to revert back to an early differentiation state. Starting from that state keratinocyte differentiation can be induced again by 2.8 mM $CaCl_2$ and the process of differentiation can be monitored over several subsequent days [129].

To corroborate the finding that LRRC8A modifies differentiation, HaCaT keratinocytes were first de-differentiated and then induced to differentiate by elevation of extracellular Ca^{2+} concentration. HaCaT cells were seeded in medium containing 0.03 mM $CaCl_2$ before medium was changed to 2.8 mM $CaCl_2$ the next day. Morphology was monitored by microscopy and pictures were taken every two days (Figure 5-23). HaCaT cells in medium containing 0.03 mM $CaCl_2$ (d0) appeared spherical, seemed to be bulged with high cell volume, nucleus and cytoplasm were distinguishable and cells grew in loose colonies that spread over the culture plate. Already one day after medium change (d1) cells started to alter their morphology and appeared smaller, flattened and edged. Five and seven days after induction HaCaT cells grew in tightly packed colonies that did not spread into the free space of the plate, but built a rim at the colonies' boarder. To control that morphological changes were induced by elevated $CaCl_2$ concentration, HaCaT cells were grown for seven days in medium containing only 0.03 mM $CaCl_2$. Cells became a little smaller and colonies grew tighter compared to d0 before induction. However, the extent of changes was not as strong as after seven days in medium containing 2.8 mM $CaCl_2$. This control experiment showed that the confluence after cultivation for seven days induced only mild morphological changes. Comparing the morphology of HaCaT-WT (Figure 5-23 A) and HaCaT-$LRRC8A^{-/-}$ cells (Figure 5-23 B) demonstrated that flattening and shrinkage of cells began earlier and tighter packed colonies were seen after shorter cultivation times for HaCaT-$LRRC8A^{-/-}$ cells. Thus, morphological changes, indicative of differentiation, appeared earlier in the absence of LRRC8A.

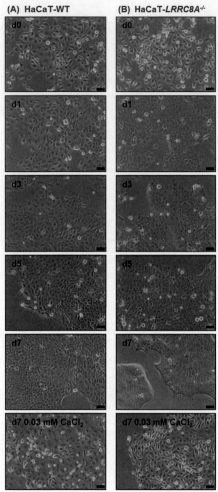

Figure 5-23: Microscopic images of HaCaT-WT and HaCaT-LRRC8A$^{-/-}$ cells during Ca^{2+}-induced differentiation

HaCaT-WT (A) and HaCaT-*LRRC8A*$^{-/-}$ cells (B) were seeded in medium with 0.03 mM CaCl2 (d0). Medium was changed to 2.8 mM CaCl2 and pictures were taken after one, three, five and seven days (d1, d3, d5 and d7). As control, cells were grown in medium containing 0.03 mM CaCl2 for seven days. Cells became smaller and grew in tighter packed colonies during cultivation in medium containing 2.8 mM CaCl2. Changes in cell morphology occurred earlier and stronger in HaCaT-*LRRC8A*$^{-/-}$ cells. Bars represent 50 μm.

Elevation of extracellular Ca^{2+} concentration resulted in visible changes of HaCaT cell morphology. To evaluate whether these morphological changes are a result of changes in differentiation state, whole cell protein lysates were prepared on seven subsequent days after induction. Involucrin and keratin 10, as markers for differentiation, were detected by

Western blot (Figure 5-24 and Supplement Figure S3). Involucrin and keratin 10 could not be detected in early differentiation state, which were achieved by cultivation of HaCaT cells in medium containing 0.03 mM $CaCl_2$. After changing to medium containing 2.8 mM $CaCl_2$ involucrin and keratin 10 increased continuously until day seven indicating ongoing differentiation of HaCaT keratinocytes. Cultivation of HaCaT cells with medium containing 0.03 mM $CaCl_2$ for seven days led to only slightly elevated involucrin and keratin 10 compared to d0, which shows that differentiation was induced by raising Ca^{2+} concentration but not high confluency during cultivation time. Importantly, involucrin and keratin 10 were detected earlier in HaCaT-$LRRC8A^{-/-}$ cells and the increase of both proteins over time was stronger compared to HaCaT-WT.

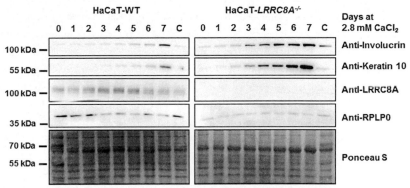

Figure 5-24: Western blot analysis of Ca^{2+}-induced differentiation of HaCaT-WT and HaCaT-$LRRC8A^{-/-}$ cells

HaCaT-WT and HaCaT-$LRRC8A^{-/-}$ cells were seeded in medium containing 0.03 mM $CaCl_2$ before changing medium to 2.8 mM $CaCl_2$ the day after seeding. As control (C), cells were grown in medium containing 0.03 mM $CaCl_2$ for seven days. Whole cell protein lysates were subjected to Western blot. Differentiation markers involucrin and keratin 10 increased over cultivation time indicating differentiation of HaCaT cells. Increase of both markers started earlier and increased stronger in HaCaT-$LRRC8A^{-/-}$ compared to HaCaT-WT cells indicating involvement of LRRC8A in HaCaT keratinocyte differentiation. Absence of LRRC8A in HaCaT-$LRRC8A^{-/-}$ cells was confirmed. Western blot analysis of RPLP0 and membrane staining with Ponceau S was used as loading control.

Taken together these results showed that HaCaT keratinocytes can be differentiated *in vitro* either by post-confluent growth or elevated Ca^{2+} concentration in the culture medium. Interestingly, it was shown that LRRC8A is involved in differentiation of HaCaT keratinocytes since in the absence of LRRC8A expression of differentiation-specific genes started at an earlier differentiation state. For that reason, it can be speculated that LRRC8A prevents premature differentiation of HaCaT keratinocytes.

5.6.3 Epidermis equivalents demonstrate disturbed differentiation of HaCaT keratinocytes in absence of LRRC8A

Here, it was shown that LRRC8A is involved in differentiation of HaCaT cells in 2D monolayer cultures. In native human epidermis keratinocytes differentiate in a basal to suprabasal direction thereby building a multilayered tissue in which keratinocytes change their morphology and gene expression profile. To expand the analysis of LRRC8A involvement in epidermal differentiation, 3D reconstructed epidermis equivalents were used. Therefore, HaCaT keratinocytes were seeded as a monolayer on a PET pore membrane, which allows supply of medium from top and bottom. By removing the medium from the top, keratinocytes were exposed to the air, which stimulates keratinocytes to start differentiation and grow in multiple layers [130].

To investigate whether LRRC8A influences differentiation in this 3D environment, the morphology of epidermis equivalents from HaCaT-WT and HaCaT-$LRRC8A^{-/-}$ cells was assessed by staining with hematoxylin and eosin. HaCaT-WT cells grew into a tightly packed epidermis equivalent, where cells stood upright in the lowest layer that was attached to the membrane. In upper cell layers, cells started to flatten and finally appeared more compact but still nucleated in the outermost cell layer, which started to shed from the epidermis equivalent. However, in contrast to native epidermis, HaCaT-WT cells did not form an anucleated *stratum corneum* (Figure 5-25 A). HaCaT-$LRRC8A^{-/-}$ cells did grow into epidermis equivalents with comparable thickness (HaCaT-WT: 104.9 ± 14.1 µm; HaCaT-$LRRC8A^{-/-}$: 109.2 ± 9.5 µm). Interestingly, morphology of HaCaT-$LRRC8A^{-/-}$ cells differed from HaCaT-WT cells. HaCaT-$LRRC8A^{-/-}$ cells appeared more voluminous and the epidermis equivalent was less tightly packed showing signs of spongiosis. Only the outermost cell layer appeared flattened and more rigid (Figure 5-25 B). Over all, morphology of reconstructed epidermis equivalents was clearly different in the absence of LRRC8A.

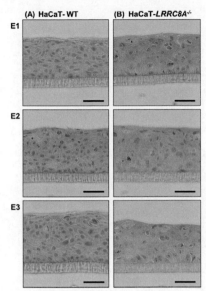

Figure 5-25: Morphology of epidermis equivalents from HaCaT-WT and HaCaT-LRRC8A$^{-/-}$ cells

For reconstructed epidermis equivalents, HaCaT-WT (A) and HaCaT-*LRRC8A$^{-/-}$* cells (B) were seeded on PET pore membranes under submerged condition before the cells were exposed to the air on top but with medium supply from below. Epidermis equivalents were stained with hematoxylin and eosin to assess morphology. HaCaT-WT cells grew into tightly packed epidermis equivalents, but epidermis equivalents from HaCaT-*LRRC8A$^{-/-}$* cells showed signs of spongiosis. The figure shows three independently grown epidermis equivalents (E 1, 2 and 3) from HaCaT-WT and HaCaT-*LRRC8A$^{-/-}$* cells. Bars represent 50 μm.

Next, it was investigated whether impaired differentiation might be the reason for the altered morphology of the reconstructed epidermis equivalents grown from HaCaT-*LRRC8A$^{-/-}$* cells. To analyze differentiation, immunohistochemical staining for the differentiation markers keratin 10 and involucrin was performed. Nuclei were counterstained with hematoxylin. Isotype control antibodies for keratin 10 and involucrin did not show any staining on the epidermis equivalents (Figure 5-26 top row) confirming specific visualization of keratin 10 and involucrin by chosen antibodies. Keratin 10 and involucrin were detected in the upper half of HaCaT-WT epidermis equivalents but not in basal and first suprabasal layers (Figure 5-26 A and C). Since the distribution of keratin 10 and involucrin in upper cell layers resembles the location of these proteins in native epidermis, HaCaT cells in epidermis equivalents are able to mimic epidermal differentiation. Strikingly, in HaCaT-*LRRC8A$^{-/-}$* epidermis equivalents keratin 10 (Figure 5-26 B) and involucrin (Figure 5-26 D) showed a disturbed distribution. Both proteins were detected in all epidermal layers except the *stratum basale*. This observation showed

that in the absence of LRRC8A, differentiation of HaCaT keratinocytes is disturbed and
seems to occur prematurely.

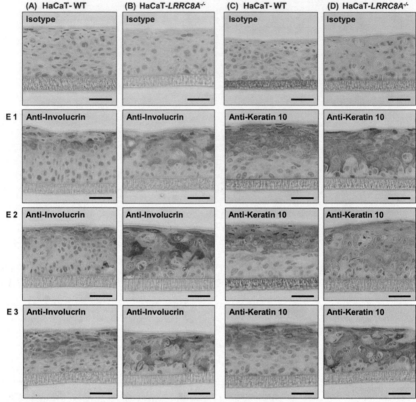

Figure 5-26: Immunohistochemical staining of epidermis equivalents from HaCaT-WT and HaCaT-LRRC8A⁻ᐟ⁻ cells

HaCaT-WT (A, C) and HaCaT-*LRRC8A⁻ᐟ⁻* cells (B, D) were used to grow epidermis equivalents that were either incubated with Anti-Keratin 10 (A, B), Anti-Involucrin (C, D) or isotype control antibodies (A-D top row). Expression of *involucrin* and *keratin 10* in the upper half of HaCaT-WT epidermis equivalents resembles the distribution of these proteins in native epidermis. In HaCaT-*LRRC8A⁻ᐟ⁻* epidermis equivalents distribution is disturbed indicating the involvement of LRRC8A in HaCaT cell differentiation. The figure shows three independently grown epidermis equivalents (E 1, 2 and 3) from HaCaT-WT and HaCaT-*LRRC8A⁻ᐟ⁻* cells. Bars represent 50 μm.

5.7 LRRC8A is not involved in barrier function of epidermis equivalents

As the outermost layer, the human epidermis is an important barrier and protects the human body against harmful environmental conditions. Most important for this protective function are the *stratum corneum*, consisting of dead corneocytes in a lipid matrix, and tight junction proteins that connect the keratinocytes of the *stratum granulosum* [33]. The proper differentiation of keratinocytes within the epidermis is essential for maintaining this important barrier function and since differentiation was altered in the absence of LRRC8A it was investigated whether barrier function is also disturbed. Therefore, epidermis equivalents were grown from HaCaT-WT and HaCaT-*LRRC8A⁻/⁻* cells and the outside-in barrier function was assessed by measuring penetration of the fluorescent dye Lucifer yellow. Lucifer yellow was applied on top of the epidermis equivalents and incubated for up to 3 h. Penetration was determined by 1) fluorometric measurement of Lucifer yellow dye that passed through the living epidermis model and 2) microscopy of fixed epidermis equivalent slides to assess the distribution of infiltrated dye. As control, SDS was applied on top of the epidermis equivalent prior to measurement to destroy the barrier and allow maximal penetration of Lucifer yellow.

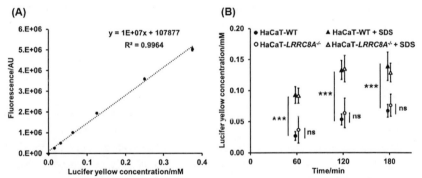

Figure 5-27: *Outside-in barrier function of HaCaT-WT and HaCaT-LRRC8A⁻/⁻ epidermis equivalents*

HaCaT-WT and HaCaT-*LRRC8A⁻/⁻* cells were used to grow epidermis equivalents to assess outside-in barrier function by flow-through of Lucifer yellow. (A) A serial dilution of Lucifer yellow solution was prepared and fluorescence was measured to generate a standard curve by plotting fluorescence signal over Lucifer yellow concentration. The graph depicts one representative example showing the mean with SD of three technical replicates and a linear regression line. (B) Basolateral Lucifer yellow fluorescence was measured 60, 120 and 180 min after application of Lucifer yellow and concentration was calculated using the standard curve. Lucifer yellow penetration increased over time, but was equal for HaCaT-WT and HaCaT-*LRRC8A⁻/⁻* cells. The graph depicts the mean with SD of at least 16 measurements from six independent experiments.

A standard curve in the range from 0.01 – 0.38 mM showed a linear relationship between fluorescence and concentration of Lucifer yellow allowing the calculation of Lucifer yellow concentration from measured fluorescence (Figure 5-27 A). The fluorescence of Lucifer yellow dye, which penetrated through the whole epidermis equivalents into the lower compartment that was filled with PBS, was measured 60, 120 and 180 min after application. SDS pretreatment of epidermis equivalents led to high flow-through of Lucifer yellow that was maximal 120 min after application. No difference between HaCaT-WT and HaCaT-*LRRC8A*$^{-/-}$ cells was observed. Penetration of Lucifer yellow through untreated epidermis equivalents was lower but increased continuously over time. Interestingly, in the absence of LRRC8A Lucifer yellow penetration was not altered (Figure 5-27 B). This finding was surprising since differentiation of HaCaT-*LRRC8A*$^{-/-}$ cells was clearly disturbed (Figure 5-26). Taken together, these results indicate that HaCaT epidermis equivalents build an outside-in barrier that can be destroyed by SDS but which is not reduced in the absence of LRRC8A despite the impact of LRRC8A on differentiation.

To evaluate if Lucifer yellow penetrated equally through the epidermis equivalents or if it was retained at a specific keratinocyte layer, the epidermis equivalents were fixed after 180 min incubation with Lucifer yellow and processed for histological analysis. Nuclei were counterstained with DAPI. In HaCaT-WT (Figure 5-28 A) as well as HaCaT-*LRRC8A*$^{-/-}$ (Figure 5-28 C) epidermis equivalents the Lucifer yellow fluorescence was detected along the smooth edge of the uppermost cell layers and weaker throughout the whole model. This fluorescence distribution visualized a functional barrier of HaCaT-WT and HaCaT-*LRRC8A*$^{-/-}$ epidermis equivalents, along which Lucifer yellow dye was retained. Additionally, green spots were detected in the epidermis equivalents that occurred more often in HaCaT-*LRRC8A*$^{-/-}$. These spots might be caused by infiltration of Lucifer yellow dye into spaces between single cells. After SDS treatment the uppermost cell layers appeared rough and the green fluorescence was detected throughout the whole epidermis equivalents (Figure 5-28 B and D) showing that Lucifer yellow could penetrate into the whole epidermis equivalent after SDS-induced barrier disruption. No obvious difference in outside-in barrier function was observed between HaCaT-WT and HaCaT-*LRRC8A*$^{-/-}$ epidermis equivalents. Furthermore, microscopy displayed that the uppermost cell layer, that can be regarded as a *stratum corneum*-like structure, was equally formed in the presence or absence of LRRC8A.

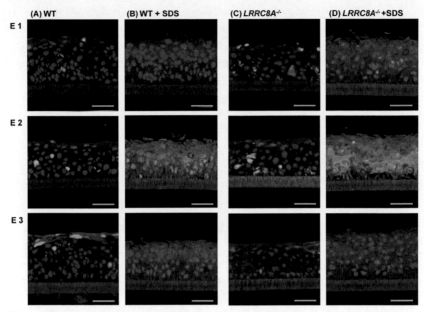

Figure 5-28: Lucifer yellow penetration in epidermis equivalents from HaCaT-WT and HaCaT-LRRC8A$^{-/-}$cells

Epidermis equivalents were grown from HaCaT-WT (A, B) and HaCaT-*LRRC8A*$^{-/-}$ cells (C, D) and pretreated with SDS (B, D) prior to incubation with Lucifer yellow. Epidermis equivalents were processed for histological analysis and nuclei were counterstained with DAPI. Lucifer yellow was restrained at the uppermost cell layer of untreated epidermis equivalents (A, C), however could enter the epidermis equivalent after SDS-induced barrier disruption (B, D). The figure shows three independently grown epidermis equivalents (E 1, 2 and 3) from HaCaT-WT and HaCaT-*LRRC8A*$^{-/-}$ cells. A FITC filter was used to visualize Lucifer yellow fluorescence. Bars represent 50 µm.

Besides the *stratum corneum*, which is only partially mimicked by HaCaT keratinocytes, tight junctions are important for the barrier property of the epidermis. Tight junctions consist of various proteins that are responsible for cell-cell contact of keratinocytes in the *stratum granulosum* thereby forming a compact and dense cell layer contributing to the epidermal barrier function. Prominent members of tight junctions are claudins as well as tight junction proteins, but also cingulin and occludin have already been investigated in human epidermis [131].

Thus, the final question that arose was whether absence of LRRC8A changes expression of genes that code for tight junctions. Therefore, HaCaT-WT and HaCaT-*LRRC8A*$^{-/-}$ cells were differentiated by post-confluent growth and RNA of early ($0.1*10^6$ cells) and late ($1.0*10^6$ cells) differentiation state was subjected to RNA sequencing. *Claudin-1* (*CLDN1*) increased from early to late differentiation state of HaCaT-WT and HaCaT-*LRRC8A*$^{-/-}$ cells while *claudin-4* (*CLDN4*) and *tight junction protein-1* and *-2* (*TJP-1* and *-2*) decreased during differentiation. However, no significant difference between HaCaT-WT and HaCaT-*LRRC8A*$^{-/-}$ cells was observed. Neither *cingulin* (*CGN*) nor *occludin* (*OCLN*) did change during differentiation of HaCaT-WT and HaCaT-*LRRC8A*$^{-/-}$ cells (Figure 5-29). Altogether, expression of typical genes encoding for tight junctions were not altered in the absence of LRRC8A.

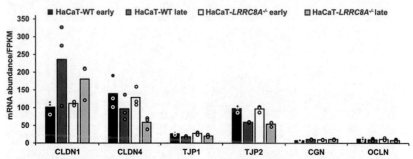

Figure 5-29: Expression of tight junctions during differentiation of HaCaT-WT and HaCaT-LRRC8A $^{-/-}$ ***cells by post-confluent growth***

HaCaT-WT and HaCaT-*LRRC8A*$^{-/-}$ cells were differentiated by post-confluent growth and RNA of earliest and latest differentiation state was subjected to RNA sequencing. mRNA abundances are displayed in fragments per kilobase of exon per million fragments mapped (FPKM) as the mean (bars) of the single values (dots) of three independent RNA preparations. The following prominent members of tight junctions were analyzed: *claudin-1 and -4* (*CLDN1* and *-4*), *tight junction protein-1 and -2* (*TJP1* and *-2*), *cingulin* (*CGN*) and *occludin* (*OCLN*). mRNA abundances were not altered in the absence of LRRC8A.

Combining the results obtained from differentiation of HaCaT keratinocyte induced by post-confluent growth and Ca^{2+} as well as in epidermis equivalent models, it was shown that differentiation-specific genes are expressed at an earlier differentiation state or in lower epidermal layers as well as in higher mRNA abundances in HaCaT-*LRRC8A*$^{-/-}$ cells, which suggests that premature and enhanced differentiation occurs in HaCaT keratinocytes in the absence of LRRC8A. However, the disturbed differentiation did not alter barrier function of HaCaT epidermis equivalents suggesting that LRRC8A is only involved in the transition from proliferation to differentiation of keratinocytes, but not for further maturation during differentiation.

6   Discussion

6.1   Keratinocyte-specific expression of the *LRRC8* gene family

The aim of this study was to investigate the expression and function of the recently discovered VRAC component LRRC8A in human keratinocytes. Phylogenetic analysis suggested an ubiquitous expression of *LRRC8A* [90] that was confirmed in a great variety of cell cultures and tissues, notably not including keratinocytes [88, 113]. In this study, it was shown that *LRRC8A* is indeed expressed in native human epidermis as well as cultured keratinocytes. Earlier studies using qRT-PCR with human *LRRC8A*-specific primers showed moderate expression of *LRRC8A* in mouse whole skin [88, 106]. However, it has to be considered that skin not only consists of keratinocytes in the epidermis, but additionally contains other epidermal, dermal and subcutaneous cell types like melanocytes, fibroblasts, sweat and sebaceous glands or adipocytes. Therefore, by using whole skin, these earlier studies did not allow conclusions about *LRRC8A* expression in a specific cell type of the various different cell types found in the skin.

Moreover, a quantitative analysis comparing the mRNA abundance of all five *LRRC8* gene family members in human keratinocytes has not been presented before. Therefore, mRNA abundance of the *LRRC8* gene family in the keratinocyte cell line HaCaT and primary human keratinocytes (NHEK) was analyzed by RNA sequencing. Highest mRNA abundance of LRRC8A was shown for HaCaT and NHEK cells as well as a distinct distribution of LRRC8A-E that was different from the kidney cell line HEK293. Different expression patterns of the LRRC8 family assessed by qRT-PCR and Western blot were also described in rat astrocytes [97] as well as mouse brain, lung and testis [109] demonstrating that the composition of LRRC8 subunits is cell-type dependent. Furthermore, LRRC8 subunits can assemble into differently composed LRRC8 hexamers and not only one but multiple different hexamers can be formed in the same cell type [103]. Since subunit composition determines VRAC inactivation kinetics [89], single-channel conductance [104], selectivity of cisplatin and taurine uptake [92] and transport of osmolytes and neurotransmitters [53], it can be speculated that different LRRC8 complexes might orchestrate cell type-specific functions also in keratinocytes.

6.2   LRRC8A is an essential VRAC component and contributes to RVD of cultured human keratinocytes

Before the molecular identification of VRACs, channel properties were assessed using pharmacological inhibitors against chloride channels and a variety of substances were tested on different cell types by diverse methods. However, a study using known chloride channel inhibitors to correlate VRAC activity with regulatory volume decrease in keratinocytes is missing. In this study, the applied inhibitors demonstrated different efficiencies of blocking VRAC activity and RVD. CBX was the most efficient inhibitor and almost completely blocked iodide influx rate and strongly reduced RVD of HaCaT keratinocytes. A similar efficiency in inhibition of VRAC currents was found in cortical rat astrocytes although the authors did not monitor RVD [132]. CBX also inhibits Ca^{2+} influx through TRPV4 channels [133], which might affect the Ca^{2+}-dependent portion of RVD as discussed later. DCPIB was described as the most selective inhibitor for VRACs [115] and yielded strong inhibition of iodide influx rate and subsequent RVD. When first described, DCPIB did not inhibit the chloride channel CFTR and different members of the CLC family when expressed in *Xenopus* oocytes nor native potassium channels in guinea-pig cardiomyoctes [115]. However, subsequent studies using DCPIB discovered inhibition of inwardly rectifying K^+ channels (K_{ir}) [134], some subunits of the two-pore-domain potassium channels (K2P) [135] or the $H^+/K^+/ATPase$ [136]. Although DCPIB inhibits different targets, Kern *et al.* showed that DCPIB occupies the selectivity filter of LRRC8A channels by electrostatic interaction [102]. The chloride channel inhibitor DIDS inhibited iodide influx rate and quenching of calcein fluorescence measured in this study to assess VRAC activity and speed of RVD. This confirmed an earlier report where DIDS mediated inhibition of VRAC activity and RVD of HaCaT cells, which were measured by patch clamp and flow cytometry, respectively [61] showing that different methods are valid to study the same phenomena. To determine whether Calcium-activated chloride channels (CaCCs) might be involved, the CaCC inhibitor NFA [116] was used, which blocked neither iodide influx nor calcein quenching indicating that CaCCs are rather not involved in RVD. This observation is different from an earlier finding, which showed that NFA inhibited VRAC currents and RVD by 20 % in HEK293 cells [137]. In summary, VRAC inhibition correlated with inhibition of RVD, which indicates that activation of VRACs contributes to cell volume decrease during hypotonic stress response of HaCaT keratinocytes. However, none of the tested inhibitors is exclusively specific for VRACs thereby hindering the clarification of VRAC involvement in RVD. Definitive statements can only be made using knock-out studies targeting the defined candidate ion channel.

The recent discovery of LRRC8A being the essential VRAC component in different cell types [88, 89] made it now possible to investigate the molecular identity of VRACs also in keratinocytes. Using the CRISPR-Cas9 technology, a monoclonal HaCaT cell line devoid of LRRC8A was generated. Strikingly, VRAC activity was nearly abolished in these HaCaT cells. This finding was corroborated by studies using primary keratinocytes, which showed a clear reduction of VRAC activity although residual LRRC8A protein was detected due to the impossibility to raise single cell clones from primary cells. The expression of *LRRC8A* as well as essential contribution to VRAC currents was confirmed for a variety of cells including HEK293 [89], HeLa [88], immune cells [88, 91], cancer cell lines [89, 92, 93], adipocytes [94], β-cells [95] and astrocytes [96]. However, for some cell types other proteins were discovered as VRAC mediators. Bestrophin1, but not LRRC8A was described as VRAC in human retinal pigment epithelium cells [138]. In gastric cancer cells [78] and astrocytes [79] the TTYH family was discovered to be essential in mediating VRAC currents. RNA sequencing in this study showed that bestrophin 1 and TTYH1 were not expressed in HaCaT and NHEK cells (Supplement Figure S4) thereby ruling out their involvement in VRAC activity. Only low and medium mRNA abundances were detected for TTYH2 and TTYH3, which were proposed as Calcium-activated chloride channels [87]. It was suggested that TTYH is activated by cell swelling and LRRC8A is activated by reduction in cytosolic ionic strength [79]. However, the role of ionic strength in activating LRRC8A is controversially discussed [104, 105, 139]. Due to the discovery of different proteins as VRAC mediators in different cell types, it can be speculated that VRACs are not only cell type-specific compositions of different LRRC8 subunits, but probably channels that are formed by different proteins in cell type-specific manner.

Irrespective of the emerging discussion about the molecular identity of VRACs in other cell types, it was shown in this study that LRRC8A is activated and furthermore mediates cell volume decrease during hypotonic stress response of keratinocytes. Earlier studies already showed that cultured keratinocytes respond to hypotonic stimulation with chloride efflux [60, 61, 140]. However, conflicting reports were published regarding the extend of RVD in keratinocytes. Cell volume of primary keratinocytes increased after hypotonic stimulation but did not decrease to the initial volume, which was explained by the lack of K^+ efflux [140]. Monitoring volume changes of HaCaT cells showed contradicting results in different studies; either increase and complete decrease to initial volume [61] or no major cell volume changes at all although VRAC currents were detected [60]. These differences can be explained by divergent measuring methods that were used in the different studies. In this study cell volume increase and decrease of HaCaT cells as well as primary keratinocytes could be demonstrated and it was shown that RVD involves LRRC8A.

Although cell volume decrease was slower in the absence of LRRC8A, keratinocytes were still able to undergo RVD. This observation differs from the initial description of LRRC8A in HEK293 cells, which could not regulate their volume in the absence of LRRC8A [89]. Nevertheless, the observation of RVD in keratinocytes devoid of LRRC8A is not surprising since RVD not solely depends on chloride efflux but also includes efflux of potassium ions and osmolytes. It was shown that different osmolytes could leave the cell through differently composed LRRC8 channels [92, 103] that might be compromised in the absence of LRRC8A. It would be interesting to investigate whether LRRC8 channels without LRRC8A are sufficient to release osmolytes from keratinocytes. Furthermore, identification of osmolyte efflux pathways independent from LRRC8 would supplement the understanding of regulatory volume decrease.

6.3 Elevated intracellular Ca^{2+} concentration increases VRAC activity and RVD during hypotonic stress response

Ca^{2+} is important for many cellular signaling cascades and considered to be involved in hypotonic stress response [60]. In this study, hypotonic stimulation of HaCaT keratinocytes resulted in increase of the intracellular Ca^{2+} concentration, which depends on the presence of extracellular Ca^{2+}. This increase of intracellular Ca^{2+} accompanied an enhancement of VRAC activity and RVD, albeit both processes did not require Ca^{2+}.

An increase of intracellular Ca^{2+} concentration upon hypotonic stimulation, shown here, was already described in earlier studies of human keratinocytes [59, 60], human lung carcinoma cells [120] or murine neuroblastoma cells [121]. It can be speculated that the intracellular Ca^{2+} concentration increases due to Ca^{2+}-induced Ca^{2+} release from intracellular stores as at least traces of Ca^{2+} were required to increase the intracellular Ca^{2+} concentration. This suggestion is in line with a study conducted in retinal Müller cells [141] and also indirectly supported by the finding that hypotonicity-induced chloride currents strongly depend on Ca^{2+} release from the endoplasmatic reticulum (ER) [142]. Moreover, a high extracellular Ca^{2+} concentration enhanced initial Ca^{2+} influx and prolonged the time of elevated Ca^{2+} concentration probably by store-operated Ca^{2+} entry (SOCE). SOCE generally proceeds after depletion of the ER-Ca^{2+} store and can only occur in the presence of extracellular Ca^{2+} that is necessary for influx into the cell and replenishment of intracellular Ca^{2+} stores [143]. Activation of SOCE upon hypotonic stimulation of keratinocytes was already proposed in an earlier report [59] strengthening the suggestion of store-operated calcium entry in hypotonic stress response of keratinocytes made in the present study.

Parallel to hypotonicity-induced increase of the intracellular Ca^{2+} concentration, also iodide influx rate increased in dependence of the extracellular Ca^{2+} concentration. Generally, VRACs are considered to act independently from Ca^{2+}; VRAC currents were still detectable with chelated intracellular Ca^{2+} in pulmonary epithelial cells [120] and increase of intracellular Ca^{2+} concentration alone is not sufficient to activate VRACs [61, 142] although a low nanomolar intracellular Ca^{2+} concentration is necessary [119, 142]. In the present investigation iodide influx was measurable in a Ca^{2+}-free environment and was increased by additional Ca^{2+}-dependent processes. It was speculated that the availability of LRRC8 complexes at the plasma membrane is increased by unfolding of caveolae upon hypotonic stimulation [144]. Furthermore, computational analysis of the LRRC8A protein sequence predicted various possible domains for calmodulin binding that might influence LRRC8A activation or modulation in a Ca^{2+}-dependent manner [120]. Looking not only on a single protein but also on the whole cell that responds to hypotonic stress a more complex picture emerges. VRACs, identified as LRRC8A complexes, and CaCCs, mainly regarded as TMEM16A, are separate chloride channel subtypes that were recently described as distinct components of swelling-induced chloride currents [142].

Co-immunoprecipitation showed that TMEM16A and LRRC8A are interacting [144] however it is assumed that both proteins form independent channels with parallel [144] or sequential [61] activation by swelling and local Ca^{2+} [142]. A possible interaction pathway was described in primary keratinocytes. It was shown that hypotonic stimulation resulted in release of ATP that can bind to purinergic receptors. In turn phospholipase C is activated, leading to an increase in intracellular Ca^{2+} concentration and subsequent activation of CaCCs [59]. Therefore, CaCCs seem to contribute to hypotonic stress response, although LRRC8A is considered to be the major component in hypotonicity-induced chloride secretion.

An increase in extracellular Ca^{2+} concentration not only correlated with increasing VRAC activity, but also with increasing RVD. Both processes, VRAC activity and RVD, acted independently from Ca^{2+}, but were supported by increased extra- and intracellular Ca^{2+}. Involvement of Ca^{2+} in RVD was reported earlier since chelation of intra- and extracellular Ca^{2+} attenuated RVD [120, 121]. Besides hypotonicity-induced chloride currents that are discussed above, further processes are involved in RVD including activation of K^+ channels or the release of osmolytes, some of which might be Ca^{2+}-dependent. Indeed, taurine but not glutamate efflux was Ca^{2+}-dependent in human retinal Müller cells [141]. Since it was shown that different osmolytes flow through differently composed LRRC8 heteromers, it can be suggested that specific LRRC8 heteromers are modulated by Ca^{2+} while others are not [141]. It will be interesting to investigate the efflux of various osmolytes through differently composed LRRC8 channels in keratinocytes and how Ca^{2+} modulates this efflux.

6.4 LRRC8A might contribute to the proliferation-differentiation transition of keratinocytes

The preferential expression of *LRRC8A* in the basal layer of native epidermis leads to different questions: 1) How do suprabasal keratinocytes regulate their cell volume when not expressing LRRC8A? 2) Is LRRC8A involved in proliferation of basal keratinocytes? 3) Is LRRC8A involved in the commitment of keratinocytes to terminal differentiation?

Cell volume regulation of suprabasal keratinocytes is particularly interesting since entry of freshwater through a disturbed epidermal barrier frequently applies hypotonic stress on differentiated keratinocytes. The effect of hypotonic stress on cell survival differs between keratinocytes in different states of differentiation; keratinocytes at an early differentiation state are more prone to hypotonic stress than at higher differentiation state. That might be explained by formation of intra- as well as intercellular adhesion molecules, which make the cells more rigid and less flexible for cell volume changes. Moreover, a connection between cell volume regulation and the cytoskeleton was reported. The actin and tubulin cytoskeleton disintegrate and quickly rearrange upon hypotonic stress while only the intermediate filament network remains stable, possibly as scaffold for initial cell shape that has to be reestablished [145]. Additionally, disruption of the fibrillary actin bundles before hypotonic stimulation of HaCaT cells resulted in elevated cell volume increase and abolished RVD indicating the importance of an intact actin cytoskeleton for RVD. In contrast to the first study, no changes in microscopic actin organization were monitored. However, *de novo* synthesis of actin indicates reorganization of the actin network. Thus, a direct interaction of actin filaments with cell volume regulation was proposed [146], probably contributing to cell volume regulation of differentiated keratinocytes.

The involvement of VRACs in keratinocyte proliferation cannot only be suggested by the expression of *LRRC8A* in basal keratinocytes, but also based on reports showing reduced cell proliferation upon VRAC inhibition [147, 148] and varying volume-sensitive chloride currents in different stages of the cell cycle [68, 149]. In this study, fewer HaCaT-*LRRC8A*$^{-/-}$ cells were counted once the cells enter the exponential growth phase and a reduced growth rate of HaCaT-*LRRC8A*$^{-/-}$ compared to HaCaT-WT cells was determined. Additionally, metabolic activity and DNA replication, as indirect measures for proliferation, were reduced in the absence of LRRC8A indicating that knock-out of *LRRC8A* resulted in reduced proliferation of HaCaT keratinocytes. Conflicting reports concerning the role of VRACs and specifically LRRC8A in proliferation of a variety of cell lines were published. DCPIB and NPPB reduced proliferation of rat astrocytes [147] and murine mesenchymal stem cells [148], respectively. Human glioblastoma cells showed lower proliferation by inhibition with DCPIB [150] and downregulation of LRRC8A [112]. Additionally, knock-down of LRRC8A in brain smooth muscle cells inhibited angiotensin II-induced proliferation via the PI3K/Akt signaling pathway [111]. In contrast, Liu and Stauber could not confirm the proliferation deficit of glioblastoma cells, myoblasts and colorectal cancer cells neither by pharmacological inhibition nor after LRRC8A knock-down [118, 151]. It has to be mentioned that the applied inhibitors are not exclusively specific for VRACs, but block chloride channels with different efficiency [137]. Therefore, an involvement of chloride channels in proliferation can be considered as highly probable although these channels might not exclusively be VRACs but rather a combination of different chloride conducting channels.

To further elucidate the role of the LRRC8 family in keratinocyte proliferation, studies should be expanded to the three keratinocyte subpopulations, which have different proliferative potentials: keratinocyte stem cells, transient amplifying cells and post mitotic cells [6]. Keratinocyte stem cells (KSCs) provide the stem cell reservoir for epidermal renewal and divide occasionally to give rise to transient amplifying cells (TAs). TAs are regarded as progenitor cells of differentiated keratinocytes and quickly undergo multiple divisions. Once TAs leave the basal layer, they are called post mitotic cells (PMs). PMs do not proliferate, but are irreversibly committed to terminal differentiation and undergo multiple changes in morphology and gene expression until finally being shed from the epidermis as corneocytes [7]. KSCs, TAs and PMs have different proliferative potentials and determining the LRRC8 subunit composition in the different keratinocyte subpopulations might give information whether one LRRC8 subunit or a certain LRRC8 subunit composition is promoting or inhibiting proliferation. Furthermore, measuring VRAC activity in KSCs, TAs and PMs might reveal whether VRAC activity acts as an additional level of regulation contributing to the multifactorial event of proliferation-differentiation transition. Indeed, VRAC currents were already linked to the transition from proliferation to differentiation of liver and muscle cells, since hypotonicity-induced chloride currents were detected in dividing hepatocytes and murine myoblasts, but not in differentiated, non-dividing hepatocytes and myotubes [152, 153]. Whether downregulation of VRAC activity is a consequence of or prerequisite for differentiation is unclear.

Finally, it was asked if LRRC8A affects keratinocyte differentiation. Earlier morphological changes, increased mRNA abundance of various typical differentiation markers as well as increased protein abundance of keratin 10 and involucrin in HaCaT-*LRRC8A*$^{-/-}$ cells compared to HaCaT-WT demonstrate the involvement of LRRC8A in keratinocyte differentiation. In epidermis equivalents as 3D model, expression of differentiation markers keratin 10 and involucrin was shown in the upper half of epidermis equivalents from HaCaT-WT cells, but already in deeper cell layers of epidermis equivalents from HaCaT-*LRRC8A*$^{-/-}$ cells. Together, these results show an altered differentiation of HaCaT keratinocytes and premature expression of differentiation markers in the absence of LRRC8A.

Unlike native epidermal keratinocytes or isolated primary keratinocytes that, once induced, stop proliferation and commit irreversibly to differentiation, HaCaT cells proliferate and differentiate at the same time and can convert between basal-like and differentiated status [129, 154]. Expression of differentiation markers did not correlate with growth arrest, but *keratin 1* and *keratin 10* were expressed in proliferating HaCaT cells [154]. Proliferation of HaCaT cells even increased upon Ca^{2+}-induced differentiation [155], which might explain the increase in *keratin 5* and *keratin 14* from early to late differentiation state seen here. Concurrent proliferation and differentiation can be explained by activation of the keratinocyte growth factor receptor (FGFR2) that is upregulated when HaCaT cells are induced to differentiate. Additionally to promoting cellular growth, FGFR2 might also be involved in signaling during differentiation [156]. In line with this report, RNA sequencing in this study showed that mRNA abundance of FGFR2 is increased from early to late differentiation status in HaCaT-WT cells (Supplement Figure S5). Furthermore, FGFR2 is downregulated in HaCaT-*LRRC8A*$^{-/-}$ cells compared to HaCaT-WT, which is conclusive with reduced proliferation in the absence of LRRC8A.

Differentiation state of keratinocytes was linked to keratinocytes size in different studies. Keratinocytes in native epidermis increase in size when switching from basal to suprabasal layers and cell size dependent expression of *involucrin* was observed in isolated stratified epithelia of different body sites [157]. Keratinocytes were separated by cell size using density gradient centrifugation and expression of *involucrin* was shown in larger (diameter >17 µm) rather than smaller keratinocytes (diameter <14 µm) [158]. Furthermore, smaller keratinocytes displayed DNA synthesis while larger keratinocytes did not [159]. This in turn supports the finding of reduced DNA synthesis of HaCaT-*LRRC8A*$^{-/-}$ cells, which also appear to be bigger. Together, it can be concluded that the size of keratinocytes correlates with their differentiation state. Furthermore, it was proposed that increase of keratinocyte size triggers terminal differentiation. One explanation for this hypothesis might be endoreplication where differentiating keratinocytes progress through the cell cycle and replicate their DNA while cell division is hindered probably by physical constrains of a rigid cytoskeleton formed by keratin filaments [17, 18]. Another explanation assumes that increased cell size leads to reduced contact between basal keratinocytes and the basal membrane, which in turn reduces receptor expression and allows delamination and subsequent movement of keratinocytes into the spinous layer [160]. Additionally, a recent study showed that cell crowding after release of mechanical stretching of epidermal progenitor cell monolayers increases the expression of differentiation specific genes and number of delaminated cells [8]. Using murine mesenchymal stem cells (mMSCs) it was furthermore shown that cell volume and stem cell fate are interconnected. Modulation of cell volume influences differentiation and determines the fate of mMSCs while induction of differentiation with known chemicals or specific medium composition changes the cell volume according to the stem cells' fate [161]. Just recently, LRRC8A was connected to differentiation of murine muscle cells and shown to mediate cellular hyperpolarization, which is necessary for the differentiation of mouse myoblast into myotubes [151]. All these studies demonstrate that differentiation is connected to cell volume changes. Since LRRC8A is meditating cell volume decrease in hypotonic stress response, it seems highly probable that LRRC8A also influences differentiation by regulating cell volume of keratinocytes. Additionally, it can be suggested that absence of LRRC8A in differentiating keratinocytes in native epidermis favors differentiation since physiological cell size increase can occur undisturbed from regulating LRRC8A activation.

It was furthermore shown that cell volume changes are linked to intracellular signaling events, which are again associated with differentiation. A study on human alveolar cancer cells showed that cell swelling resulted in increased activation of Akt while lower Akt activation resulted in reduced VRAC activity [93]. A direct link between Akt signaling and LRRC8A was shown in cerebrovascular smooth muscle cells and murine thymocytes since siRNA-mediated knock-down or knock-out of *LRRC8A* led to lower Akt phosphorylation [106, 111]. Interestingly, activation of the Akt signaling pathway induced growth arrest and triggered differentiation [14, 15, 162]. Furthermore, disturbances in Akt and subsequent mTOR signaling was shown to contribute to pathogenesis of psoriasis [15]. The interaction of LRRC8A with Akt in keratinocytes and its role in epidermal homeostasis is yet unknown, although the importance of LRRC8A in skin homeostasis was already demonstrated in *lrrc8a*−/− mice, which show an aberrant skin phenotype with epidermal hyperkeratosis [106]. A comprehensive understanding of the signaling pathways remains to be elucidated to understand how LRRC8A controls the transition from proliferation to differentiation thereby contributing to the maintenance of epidermal homeostasis. From a medical perspective, it might be relevant to investigate these signaling events in the skin to further understand the pathogenesis of inflammatory dermatoses, such as psoriasis to validate LRRC8A as a novel target for the treatment of inflammatory dermatoses.

6.5 Perspective

This study was the first to investigate LRRC8A function in human keratinocytes [72]. It was shown that LRRC8A is an essential component for VRAC activity and mediates cell volume decrease in cultured human keratinocytes. It was furthermore demonstrated that proliferation is reduced and differentiation is disturbed in HaCaT cells devoid of LRRC8A. This however did not influence the barrier function of HaCaT epidermis equivalents.

The preferential localization of LRRC8A in the basal layer of human native epidermis points towards an involvement of LRRC8A in the transition from keratinocyte proliferation to differentiation during epidermal maturation and stratification. However, the precise involvement of LRRC8A in this transition remains to be determined. Investigating the LRRC8 subunit composition as well as VRAC activity and cell volume changes in keratinocyte stem cells, transient amplifying cells and post-mitotic cells might allow the correlation of LRRC8-mediated volume regulation with proliferation and differentiation of keratinocytes. It will be also interesting to understand if cell volume changes that occur during keratinocyte differentiation in native epidermis are actively regulated to control differentiation or if morphological changes are a consequence of altered gene expression. Furthermore, investigations regarding the connection of LRRC8A with signaling pathways during transition from proliferation to differentiation would be important to further understand the regulation of epidermal homeostasis. Studies conducted in physiologically healthy keratinocytes should be expanded to pathological skin conditions such as psoriasis or atopic dermatitis. This might reveal the involvement of LRRC8A in inflammatory dermatoses with disturbed epidermal homeostasis, which is of great medical interest when postulating LRRC8A as possible target for a therapeutic intervention.

7   Material and Methods

7.1   Tissue culture

7.1.1   General tissue culture conditions

All tissue culture work was performed under sterile conditions using a laminar flow cabinet. Culture medium was preheated to 37 °C in a water bath. All other solutions were used at room temperature (RT). Cells were routinely grown in 10 cm dishes and incubated at 37 °C and 5 % CO_2 in a cell culture incubator.

7.1.2   Cultivation of HaCaT cells

The spontaneously immortalized human keratinocyte cell line (HaCaT) was cultured in Dulbecco's modified eagle's medium (DMEM) with high glucose supplemented with 8 % (v/v) fetal calf serum (FCS) and 3.5 mM L-Glutamine up to maximal confluence of 80 %. For cultivation in early differentiation state, HaCaT cells were cultured in DMEM high glucose but $CaCl_2$-free medium supplemented with 10 % (v/v) FCS, 4 mM L-Glutamine and 0.03 mM $CaCl_2$. For induction of differentiation, the $CaCl_2$ concentration in the medium was raised to 2.8 mM. For passaging cells, culture medium was removed and cell monolayer was washed with 1x phosphate buffered saline (PBS). Then cells were enzymatically detached by incubation with 1 ml trysin-like enzyme (TrypLE) for 8 min in the cell culture incubator and resuspended in 9 ml medium. The cell suspension was split in a ratio of 1:8 to 1:30 in new culture dishes containing fresh culture medium and cultivated for another three to four days.

7.1.3   Cultivation of NHEK cells

Normal human epidermal keratinocytes (NHEK) were purchased from PromoCell. In this study, all experiments were performed with NHEK cells isolated from labia of a single Caucasian female donor. Cells were cultured in keratinocyte growth medium 2 (PromoCell) containing supplement mix (0.004 ml/ml bovine pituitary extract, 0.125 ng/ml epidermal growth factor, 5 µg/ml insulin, 0.33 µg/ml hydrocortisone, 0.39 µg/ml epinephrine, 10 µg/ml transferrin) and 0.06 mM $CaCl_2$ up to a confluence of 80 %. For passaging cells, culture medium was removed and cell monolayer was washed with 2 ml 2-[4-(2-hydroxyethyl)piperazin-1-yl]ethanesulfonic acid buffered balanced salt solution (HEPES BBS). Then cells were enzymatically detached by incubation with 1 ml trypsin/ethylenediaminetetraacetic acid (0.04 % (w/v) trypsin, 0.03 % (w/v) EDTA) for 10 min in the laminar flow and gentle tapping of the culture dish. Detachment was stopped with 1 ml trypsin neutralization solution (TNS) (0.05 % (w/v) trypsin inhibitor, 0.1 % (w/v) bovine serum albumin). Resuspended cells were centrifuged for 3 min at 800 rounds per minute (rpm) at RT. Supernatant was aspirated and cells were resuspended in 1 ml culture medium. Cell number and viability was determined by manual counting (see 7.1.4). Cells were seeded in new culture dishes containing fresh culture medium at a density of $0.6*10^6$ cells/10 cm dish, $0.3*10^6$ cells/10 cm dish or $0.15*10^6$ cells/10 cm dish for two, three or four days cultivation.

7.1.4 Determination of cell number

Automated Cell Counter

Electric CASY cell counter was used to automatically determine the cell number of HaCaT cells. Cells were enzymatically detached from cell culture plates as described in 7.1.2. 50 µl cell suspension was mixed with 10 ml CASY ton solution and placed in the cell counter. Determination of cell number and viability is based on electric current exclusion. Single cells pass the measuring pore and are exposed to an electric field. Since the intact plasma membrane of viable cells acts as an insulator a low current is recorded when a viable cell passes the pore whereas a higher current is measured for non-viable cells with a damaged plasma membrane. Due to the recorded currents cell viability can be accurately calculated in a dye-free process.

Neubauer Counting Chamber

A disposable Neubauer counting chamber was used to determine the cell number of NHEK and adenovirally transduced HaCaT cells. Cells were enzymatically detached from culture dishes as described in 7.1.2 and 7.1.3. Equal volumes of cell suspension and 0.04 % (w/v) trypane blue solution were mixed before 10 µl of cell suspension were pipetted into the counting chamber. Cells in the four big squares in every edge were manually counted under the microscope. The arithmetic mean was calculated to determine the cell number in 0.1 µl cell suspension.

7.1.5 Freezing and thawing of tissue culture cells

HaCaT cells were grown to a confluence of 70 % and detached as described above (see 7.1.2). Cells were centrifuged at 800 rpm for 5 min at RT and carefully resuspended in freezing medium (70 % (v/v) culture medium, 20 % (v/v) FCS, 10 % (v/v) dimethylsulfoxide (DMSO)) in a density of $0.8*10^6$ cells/ml. 1.8 ml aliquots were placed in a freezing container and stored at -80 °C overnight. For long-term storage, cells were kept in liquid nitrogen tanks.

Before thawing HaCaT cells, 5 ml culture medium were pipetted in a 10 cm dish and stored in a cell culture incubator for equilibration of temperature and pH. Frozen cells were quickly thawed in a 37 °C water bath and carefully diluted in 10 ml culture medium. Cells were centrifuged at 800 rpm for 5 min at RT, resuspended in 5 ml fresh culture medium and seeded in the prepared 10 cm dish. First passaging of cell culture was performed when cells reached a confluence of 90 %.

Before thawing NHEK cells, 9 ml culture medium were equilibrated in a cell culture incubator. Frozen cells were thawed in a 37 °C water bath and carefully transferred to the equilibrated 10 cm dish. After 24 h, culture medium was exchanged with fresh culture medium. First passaging of cell culture was performed when cells reached a confluence of 70 %.

7.1.6 Generation of *LRRC8A* knock-out cells using CRISPR-Cas9 technology

Stable HaCaT-*LRRC8A*$^{-/-}$ knock-out cell line

To raise a monoclonal HaCaT cell line carrying mutations in *LRRC8A* gene, HaCaT cells were transduced with adenovirus Ad5-CMV-Cas9-wt-2A-OFP delivering a gene expression cassette encoding for *Streptococcus pyogenes* Cas9 nuclease, fused to nuclear-localization sequences, followed by a 2A-linker and orange fluorescent protein (OFP) under the control of the cytomegalovirus (CMV) promoter and with a second adenovirus Ad5-U6-sgRNA-LRRC8A#1-U6-sgRNA-LRRC8A#2 delivering a gene expression cassette encoding two single guide RNAs (sgRNA) targeting different positions of *LRRC8A* (Table 7-8) under the control of the U6 promoter. Each adenovirus was applied with a multiplicity of infection (MOI) of 50 for 8 h with 8 µg/ml polybrene before culture medium was changed and cells were cultivated for another 72 h. Monoclonal HaCaT-*LRRC8A*$^{-/-}$ cell lines were generated by limiting dilution. Therefore, cells were diluted to statistically 1 – 4 cells/100 µl, seeded in 96-well plates and raised to monoclonal cell lines. Genomic deletion of *LRRC8A* was confirmed by target site-specific polymerase chain reaction (PCR) (see 7.2.4) and subsequent Sanger sequencing (see 7.2.6). Absence of protein was determined by Western blot analysis (see 7.2.9).

NHEK-*LRRC8A* knock-out cell populations

To generate NHEK cells carrying mutations in the *LRRC8A* gene, NHEK cells were transduced with adenoviruses Ad5-CMV-Cas9-wt and Ad5-U6-sgRNA-G-LRRC8A-#1-U6-sgRNA-G-LRRC8A-#2. Each adenovirus was applied with MOI 100 for 8 h with 8 µg/ml polybrene. Cells were used for functional analysis 72 h after transduction. Genomic deletion of *LRRC8A* was confirmed by target site-specific PCR (see 7.2.4) and subsequent nanopore sequencing (see 7.2.7). Residual protein was determined by Western blot analysis (see 7.2.9).

7.1.7 Generation of stable HaCaT-*hsYFP* cells

To generate a HaCaT cell line stably expressing the iodide sensor halide-sensitive yellow fluorescent protein (hsYFP), HaCaT cells were transduced with lentivirus CLV-CMV-YFP-IRES-Puro delivering a gene expression cassette encoding for hsYFP (YFP-H147Q/I152L/F46L) [116] under the control of the CMV promoter and a puromycin antibiotic resistance gene. HaCaT cells were transduced with MOI 20 in presence of 8 µg/ml polybrene for 8 h. Cells were selected in the presence of 0.5 µg/ml puromycin and monoclonal HaCaT-*hsYFP* cell lines were generated by limiting dilution. Monoclonal growth was confirmed by microscopy and clones were analyzed for robust hsYFP fluorescence.

7.1.8 Transduction of HaCaT cells and NHEKs to express hsYFP

To measure VRAC activity in HaCaT and NHEK wildtype and *LRRC8A* knock-out cells the hsYFP gene expression cassette is delivered via adenoviral transduction. Therefore, cells were seeded in 6-well plates (HaCaT: $0.25*10^6$ cells/well; NHEK: $0.1*10^6$ cells/well) in 2 ml culture medium and cultivated for 24 h. Then cells were transduced with adenovirus Ad5-CMV-hsYFP at MOI 300 (HaCaT) or MOI 200 (NHEK) and incubated for additional 24 h. High transduction efficiency was validated using a fluorescence microscope before cells were seeded in 96-well black-walled, clear bottom microplates (HaCaT: 30 000 cells/well; NHEK: 25 000 cells/well) and incubated for another 24 h. Cellular YFP fluorescence upon different stimuli was measured in an automated fluorescence plate reader (see 7.3.1).

7.1.9 Differentiation of HaCaT cells by Ca²⁺ induction

In order to reverse HaCaT cells to an early differentiation state, cells were cultivated in low Ca^{2+} DMEM (DMEM high glucose, $CaCl_2$-free supplemented with 10 % (v/v) FCS, 4 mM L-Glutamine and 0.03 mM $CaCl_2$) for three weeks according to protocol by Deyrieux *et al.* [129]. Early differentiation state was confirmed by reduced keratin 10 and involucrin level by Western blot analysis.

To induce differentiation, HaCaT cells at early differentiation state were detached from culture plate, counted and diluted in low Ca^{2+} DMEM to a cell density of $0.2*10^6$ cells/ml. Cell suspension was stepwise diluted 1:2 resulting in cell densities from $0.2*10^6$ cells/ml to 1560 cells/ml and 2 ml of each cell suspension was pipetted into two wells of a 6-well plate. The first plate (with a seeded cell number of $0.4*10^6$ cells/well) was used 24 h after seeding for preparation of whole cell lysates as described in 7.2.8. representing the earliest differentiation state. Culture medium in the other plates was changed to high Ca^{2+} DMEM (DMEM high glucose, $CaCl_2$-free supplemented with 10 % (v/v) FCS, 4 mM L-Glutamine and 2.8 mM $CaCl_2$). Culture medium was changed to fresh high Ca^{2+} DMEM after four and seven days. Every day one plate was used for preparation of whole cell lysates (see 7.2.8).

7.1.10 Differentiation of HaCaT cell by post-confluent growth

HaCaT cells were grown routinely, detached from the culture plate, counted, diluted in culture medium to a cell density of $1*10^6$ cells/ml and 0.1, 0.2, 0.25, 0.4, 0.6 and $1.0*10^6$ cells were seeded in 2 wells of a 6-well plate. Cells were incubated at 37 °C and 5 % CO_2 for 48 h before both wells were pooled for isolation of RNA (see 7.2.1) or preparation of whole cell lysates (see 7.2.8).

7.1.11 Generation of epidermis equivalents

To assess differentiation of HaCaT keratinocytes in a 3D environment, epidermis equivalents were grown according to the protocol by Bürger *et al.* [15]. HaCaT cells were routinely grown, detached from the culture plate, centrifuged for 5 min at 800 rpm and resuspended in CnT prime epithelial culture medium to a cell density of $1*10^6$ cells/ml. A 12-well plate was filled with 1 ml CnT prime epithelial culture medium per well and a thin cert 12-well transwell insert was placed in the 12-well plate with contact to the medium below. 500 µl cell suspension was pipetted into the transwell insert and incubated for 24 h before medium in the transwell was exchanged by fresh CnT prime epithelial culture medium to remove unattached cells. After another 48 h medium below and in the transwell insert was replaced with CnT prime 3D medium. After 24 h 4 ml CnT prime 3D medium was pipetted into a 12-well deep well plate, medium from the transwell insert was thoroughly and carefully removed and transwell insert was transferred into the deep well plate with contact to the medium below and cell exposure to the air to induce differentiation. After three, five and seven days, medium below the transwell insert was replaced by fresh CnT prime 3D medium. After another three days, epidermis equivalents were either used for determination of barrier function using Lucifer yellow (see 7.3.7) or fixed with 4.5 % (v/v) phosphate-buffered formaldehyde (PFA) overnight at 4 °C. Fixed epidermis equivalents were embedded in paraffin and stored at RT until processed for immunohistochemical staining (see 7.2.10).

7.2 Molecular biology

7.2.1 RNA isolation from tissue culture cells

Cells were seeded in 6-well plates in different cell densities (HEK293: $1*10^6$ cells/well; HaCaT: $0.5*10^6$ cells/well; NHEK: $0.25*10^6$ cells/well) and incubated for 24 h at 37 °C and 5 % CO_2. Cells were washed with PBS and scraped from the plate bottom using a cell scraper. One 6-well was used for total RNA isolation from HEK293 cells, whereas two 6-wells of the same cell type were combined for HaCaT or NHEK. Total RNA was isolated using NucleoSpin-RNA II Kit (Macherey-Nagel). In brief, cells were lysed using lysis buffer and β-mercaptoethanol and cell debris were removed by centrifugation. RNA and DNA were bound to silica membrane and DNA was digested by DNaseI. Membrane was subsequently washed and RNA was finally eluted in 60 µl RNase-free water and stored at -80 °C. RNA concentration was calculated from absorbance at 260 nm, which was measured using a NanoDrop spectrophotometer.

7.2.2 Transcriptome data analysis

RNA sequencing experiments were carried out to evaluate mRNA transcript abundances in HaCaT, NHEK and HEK293 cells. RNA isolation was performed as described in 7.2.1 and send to GATC Biotech AG or LGC Genomics for sequencing. For library preparation, TruSeq RNA Library Prep Kit v2 from Illumina starting with an input amount of 500 ng of total RNA was used. The prepared libraries were sequenced with a 2×150 bp read length using the HiSeq 3000/4000 SBS Kit and an Illumina Hiseq 4000 sequencer. The adapter trimmed, demultiplexed and quality filtered reads were aligned to the hg19 reference genome and transcriptome using Hisat2 (version 2.0.4) [163]. The Hisat2 output files

(SAM) were converted to the BAM format and were sorted and indexed using SAMtools (version 1.3.1) [164]. The further processing of the sorted BAM files was carried out using Cufflinks (version 2.1.1) to quantify the transcript abundances displayed in fragments per kilobase of exon per million fragments mapped (FPKM).

7.2.3 Isolation of genomic DNA from tissue culture cells

Genomic DNA (gDNA) was isolated by using QuickExtract DNA Extraction Solution (Epicentre) to determine genomic modifications of *LRRC8A* upon genome engineering via CRISPR-Cas9 technology. Transduced cells were analysed 72 h after transduction. HaCaT or NHEK cells were detached by incubating with Tryp/LE or Trypsin/EDTA and resuspended in culture medium or TNS, respectively. Cells were centrifuged at 800 rpm for 5 min at RT. For isolation of gDNA, cell pellet was resuspended in QuickExtract DNA Extraction Solution and heated at 68 °C for 15 min followed by 95 °C for 8 min. Extracted gDNA was stored at -20 °C.

7.2.4 Target site-specific PCR

For genome engineering HaCaT and NHEK cells were transduced with adenoviruses as described in 7.1.6. To test whether the Cas9 nuclease-induced double strand breaks led to changes in the DNA sequence by error-prone DNA repair, PCR was used to amplify DNA fragments, in which the targeted sequences are located.

Reaction was performed in a total volume of 50 µl containing 1 U Phusion DNA-polymerase, 1x high-fidelity buffer, forward and reverse DNA-primer (each 25 pmol) (see Table 7-7) and 0.2 mM dNTP. 2 µl extracted gDNA that was isolated from transduced cells (see 7.2.3) were used as DNA template. Reaction protocol is shown in Table 7-1.

Table 7-1: Reaction protocol for DNA amplification by PCR

initialisation	98 °C 30 s	
denaturation	98 °C 10 s	
annealing	65 °C 10 s	30x
elongation	72 °C 30 s	
final extension	72 °C 10 min	
hold	10 °C ∞	

7.2.5 Agarose gel electrophoresis

For visualization of DNA fragments, PCR reactions were mixed 1:6 with OrangeG loading dye (6x) and separated on 2 % (w/v) agarose gels in 1x TRIS-acetic acid-EDTA (TAE) buffer (0.4 M TRIS-acetic acid pH 8, 1 mM EDTA) containing Midori Green at 100 V. DNA was visualized under ultraviolet (UV) radiation in an Infinity Serial N gel documentation system. Sizes of DNA fragments were related to a 100 bp DNA ladder.

7.2.6   Sanger sequencing

PCR reaction mix was purified by QiaQuick PCR purification kit (Qiagen). PCR products >100 bp were bound to a silica-membrane, washed and eluted in 30 µl nuclease-free water. For sequencing at GATC Biotech AG, 2 µl purified PCR product were mixed with 25 pmol forward DNA-primer (see Table 7-7) in a final volume of 10 µl. CRISPR-Cas9-dependent genome editing was determined by comparing the genomic DNA sequences of transduced and wild-type cells using TIDE analysis [165].

7.2.7   Nanopore sequencing

To determine *LRRC8A* gene disruption in NHEK-*LRRC8A* KO populations, Nanopore sequencing (Minion, Oxford Nanopore Technologies) was used. Therefore, gDNA of each generated NHEK-*LRRC8A* KO population was amplified by target site-specific PCR (see 7.2.4). PCR amplification products were combined and 1 µg total PCR amplification product was used to prepare a 1D sequencing library (SQK-LSK109). A primed R9.4 flow cell was loaded with 75 µl of the sequencing library and ran for approximately 1 h yielding 199 247 amplicon sequences, which were base-called using Albacore (version 2.3.1). Quality scores and read lengths were visualized using NanoPlot (version 1.13.0) [166] and the alignment to the amplicon reference sequence was performed using Minimap2 (version 2.11) [167].

7.2.8   Preparation of whole cell lysates

For preparation of whole cell lysates, cells were seeded in 6-well plates (HaCaT: $0.5*10^6$ cells/well; NHEK: $0.25*10^6$ cells/well) and incubated for 24 h at 37 °C and 5 % CO_2. Cells were washed with cold 1x PBS containing protease inhibitor mix M (Serva Electrophoresis) (1:1000) and PBS was thoroughly aspirated. Whole cell lysates were prepared by incubation of cells with RIPA lysis buffer (50 mM Tris-HCl pH 7.4, 150 mM NaCl, 1 % (v/v) nonidet P-40, 0.5 % (v/v) deoxycholate sodium solution (10 % (w/v)), 0.1 % (v/v) SDS solution (20 % (w/v)), 1 mM EDTA) containing protease inhibitor mix M (1:100) for 30 min at 4 °C under gentle rocking. Supernatant was collected, centrifuged for 1 min at 4 °C and cell lysate were transferred into a cold reaction tube.

Protein concentration was determined by bicinchoninic acid (BCA) assay (Thermo Fisher Scientific). In brief, BCA reagents were mixed with cell lysate and incubated at 37 °C for 30 min. In addition, bovine serum albumin was used to obtain a standard curve in a concentration range from 2 – 0.125 mg/ml. Absorption was measured at 562 nm and protein concentration of the whole cell lysates was calculated using the linear range of the standard curve.

For SDS-polyacrylamide gel electrophoresis (SDS-PAGE) whole cell lysates were denatured by incubation with Laemmli sample buffer (0.5 M Tris-HCl pH 6.8, 4 % (w/v) SDS, 40 % (v/v) glycerol, 100 mM DTT, 0.08 % (w/v) bromophenol blue) for 10 min at 95 °C and stored at -20 °C.

7.2.9 SDS-PAGE and Western blot

Proteins of whole cell lysates were separated using 1 mm thick SDS-polyacrylamide (PAA) gels for gel electrophoresis. Gels were composed of a separating and a stacking gel, which were composed as follows: separating gel: 12.5 % (v/v) acrylamide/bisacrylamide (37.5:1) solution, 360 mM Tris-HCl pH 8.8, 0.1 % (v/v) SDS solution (20 % (w/v)), 0.11 % (v/v) TEMED, 0.1 % (v/v) ammonium persulfate solution (10 % (w/v)); stacking gel: 5 % (v/v) acrylamide/bisacrylamide (37.5:1) solution, 125 mM Tris-HCl pH 6.8, 0.1 % (v/v) SDS solution (20 % (w/v)), 0.15 % (v/v) TEMED, 0.1 % (v/v) ammonium persulfate solution (10 % (w/v)). Gels were polymerized overnight at 4 °C.

To separate proteins of whole cell lysates, 10 – 20 µg total protein were loaded on 12.5 % SDS-PAA gels and electrophoresis was performed in 1x Tris-Glycine SDS running buffer (Novex) (80 V until proteins reached the separating gel, 180 V until 35 kDa marker protein reached end of the gel). Proteins were blotted onto a PVDF membrane in 4 °C cold transfer buffer (25 mM Tris, 190 mM glycine, 20 % (v/v) methanol, 0.02 % (v/v) SDS solution (20 % (w/v))) at 350 mA for 50 min. Transferred protein was visualized by Ponceau S staining after incubation of membrane with Ponceau S solution (0.1 % (w/v) Ponceau S, 5 % (v/v) acetic acid) for 30 min and rinsing with distilled water until protein with minimal background was visible. Image of the membrane was captured using a Vilber Fusion Fx7 documentation system before membrane was cleared from Ponceau S by washing with distilled water for 5 min.

Membrane was blocked for 1 h with 5 % (w/v) skim-milk powder in Tris-buffered saline with Tween20 (TBST; 2 mM Tris-HCl pH 8.6, 150 mM NaCl, 0.05 % (v/v) Tween20). Membrane was cut into pieces to visualize proteins of different size that were separated on the same gel. Membranes were probed overnight at 4 °C with specific primary antibodies against LRRC8A (Novus Biologicals NBP2-32158 1:500), Keratin 10 (abcam ab76318, 1:10 000), Involucrin (abcam ab20202, 1:10 000), RPLP0 (abcam ab192866, 1:100), YWHAZ (Santa Cruz Biotechnology sc-293415, 1:100), β-Actin (Sigma Aldrich A1978, 1:10 000) and α-Tubulin (Sigma Aldrich T9026, 1:500). Membranes were washed three times for 10 min with TBST under gentle rocking. Membranes were incubated with species-specific, HRP-coupled secondary antibodies (α-mouse IgG or α-rabbit IgG, VWR, 1:5 000) for 1 h at RT and washed with TBST as described above. Membranes were assembled onto a plastic sheet and placed in the Vilber Fusion Fx7 documentation system. Enhanced chemiluminescence (ECL) reagents (Merck Millipore) were mixed and pipetted onto the membrane. Protein was visualized at different exposure times and images were taken using the Vilber Fusion Fx7 documentation system.

7.2.10 Immunohistochemistry of skin specimen and epidermis equivalents

Healthy volunteers gave written informed consent for harvesting of skin biopsies. The study was approved by the ethics committee of the Clinic of the Goethe-University

(116/11) and the Declaration of Helsinki protocols were followed. Punch biopsies (6 mm) were taken from healthy volunteers by trained physicians. Samples were fixed in 4.5 % (v/v) PFA and embedded in paraffin. Samples were cut into 4 μm sections using a microtome and sections were placed on a glass slide. Sections were freed from paraffin and rehydrated by serial bathing in xylol, isopropanol, 96 % (v/v) ethanol, 70 % (v/v) ethanol and distilled water. Sections were incubated at 95 °C in citrate buffer pH 6.0 (Zytomed) for involucrin antigen retrieval or Tris-EDTA buffer pH 9.0 (Zytomed) for keratin 10 and LRRC8A antigen retrieval. Slides were cooled down to RT and washed in distilled water and TBS buffer (Zytomed). Primary antibodies (Anti-LRRC8A: NBP2-32158, 1:100, Novus Biologicals; Anti-Involucrin: sc-28557, 1:3 000, Santa Cruz Biotechnology; and Anti-Keratin 10: ab76318, 1:5 000, abcam) or concentration adjusted isotype control antibody were diluted in signal stain buffer and incubated overnight at 4 °C. Sections were washed with TBS to remove primary antibodies. Antibody binding was visualized by dye precipitation , which formed after incubation with Histofine Simple Stain AP Multi (Medac Diagnostika) followed by incubation with substrate solution permanent AP red kit (Zytomed). Sections were washed with distilled water, nuclei were stained with hematoxylin and sections were dehydrated by serial incubation in 96 % (v/v) ethanol, isopropanol and xylol. Sections were mounted using enthelan. The same immunohistochemical procedure was performed on epidermis equivalents grown from HaCaT-WT and HaCaT-*LRRC8A⁻ᐟ⁻* cells.

7.3 Cell-based measurements

7.3.1 Measuring of VRAC activity using hsYFP

HaCaT and NHEK cells expressing hsYFP either due to stable integration of the corresponding gene expression cassette (see 7.1.7) or after adenoviral transduction (see 7.1.8) were seeded in 96-well black-walled, clear bottom microplates (30 000 HaCaT cells; 25 000 NHEK cells). After 24 h, cells were washed three times with 70 μl isotonic incubation buffer (see Table 7-2) and incubated with 50 μl isotonic incubation buffer for 15 min at 37 °C in an automated plate reader. Cellular hsYFP fluorescence of HaCaT-hsYFP cells with stable expression of hsYFP was continuously recorded every second in the plate reader FLIPR tetra (Molecular Devices) with excitation at 470 – 495 nm and emission at 515 – 575 nm. Adenovirally transduced cells were measured using the plate reader FLEX Station 2 (Molecular Devices) with the following settings: excitation at 485 nm, emission at 535 nm, continuous recording every 3 sec. After baseline recording for 30 sec, cells were stimulated by addition of 125 μl isotonic or hypotonic I⁻-buffer (Table 7-2) to establish an extracellular I⁻ concentration of 50 mM and in case of hypotonic stimulation a reduced osmolarity of 229 mOsm. Hypotonic buffer of different osmolarity were prepared by adjusting the NaCl concentration. Inhibitors were diluted in isotonic incubation buffer and hypotonic I⁻-buffer. Cells were incubated with inhibitor in isotonic incubation buffer for 15 min prior to measurement. For experiments without extracellular Ca^{2+}, 2 mM CaCl2 were replaced by 2 mM MgCl2 and additional 1 mM EGTA.

Table 7-2: Buffer for the measurement of VRAC activity

Isotonic incubation buffer 329 mOsm, pH 7.2	5 mM KCl 1 mM MgCl$_2$ 2 mM CaCl$_2$ 10 mM HEPES-NaOH 10 mM Glucose 145 mM NaCl
Hypotonic I⁻-buffer 189 mOsm, pH 7.2	5 mM KCl 1 mM MgCl$_2$ 2 mM CaCl$_2$ 10 mM HEPES-NaOH 10 mM Glucose 5 mM NaCl 70 mM NaI
Isotonic I⁻-buffer 329 mOsm, pH 7.2	5 mM KCl 1 mM MgCl$_2$ 2 mM CaCl$_2$ 10 mM HEPES-NaOH 10 mM Glucose 5 mM NaCl 70 mM NaI 140 mM Mannitol

HsYFP fluorescence was normalized to fluorescence value at the time point of stimulation and plotted over time. I⁻ influx rate was derived from the linear slope m calculated from the change in hsYFP fluorescence over a time frame of 100 sec ($\Delta F/\Delta t$). I⁻ influx rate was used as a measure of hypotonicity-induced VRAC activity.

To quantitatively determine the effect of inhibitors or the reduction in VRAC activity by *LRRC8A* knock-out, VRAC activity was normalized. Therefore, the gradient angle α was calculated from the slope m using the arctan function ($\alpha=\arctan(m)$). To correct for background activity, the gradient angle of isotonic control (α_{iso}) was subtracted ($\alpha_{correct}=\alpha-\alpha_{iso}$). Then calculated values were related to hypotonic control (α_{hypo}) and finally displayed in percent of hypotonic control ($\alpha_{correct}/\alpha_{hypo}*100$), which was defined as normalized I⁻ influx rate (%).

To determine the half-maximal inhibitory concentration (IC$_{50}$) of the inhibitors CBX and DCPIB, the normalized I⁻ influx rate (%) was plotted against the individual inhibitor concentration. IC$_{50}$ values were determined from curve fitting (GraFit, version 7.0.2, Erithacus Software Ltd, West Sussex, UK) using the following model: $y = a + \dfrac{b-a}{1+(\frac{x}{IC_{50}})^s}$, where y is the VRAC activity, a is the minimal and b is the maximal activity, x is the inhibitor concentration and s describes the steepness of the slope.

7.3.2 Measuring of regulatory volume decrease using calcein-AM

To measure cell volume changes, 30 000 HaCaT cells or 25 000 NHEK cells were seeded in 96-well black-walled, clear bottom microplates and cultivated for 24 h. Prior of measurement cells were loaded with 10 μM calcein-AM diluted in culture medium for 60 min in a cell culture incubator. Cells were washed three times with 70 μl isotonic incubation buffer (see Table 7-3) and incubated in 50 μl isotonic incubation buffer at 37 °C for 5 min in a fluorescence plate reader. Cellular calcein fluorescence (excitation 470 – 495 nm, emission 515 – 575 nm) was continuously recorded every second in an automated fluorescence plate reader (FLIPR Tetra, Molecular Devices). Cells that were transduced with adenovirus (see 7.1.6) prior to calcein-AM loading were measured using fluorescence plate reader FLEX Station 2 (Molecular Devices) with the following settings: excitation at 485 nm, emission at 535 nm, continuous recording every 1.6 sec.

After baseline recording for 30 sec, cells were stimulated by addition of 125 μl isotonic or hypotonic buffer (see Table 7-3). Addition of hypotonic buffer resulted in a reduction of extracellular osmolarity to 150 mOsm. Hypotonic buffer of different osmolarity were prepared by adjusting the NaCl concentration. Inhibitors were diluted in isotonic incubation buffer and hypotonic buffer. Cells were incubated with inhibitor in isotonic incubation buffer for 15 min prior to measurement. For experiments without extracellular Ca^{2+}, 2 mM $CaCl_2$ were replaced by 2 mM $MgCl_2$ and additional 1 mM EGTA.

Table 7-3: Buffers for the measurements of RVD and intracellular Ca^{2+} concentration

Isotonic incubation buffer	5 mM KCl
329 mOsm, pH 7.2	1 mM $MgCl_2$
	2 mM $CaCl_2$
	10 mM HEPES-NaOH
	10 mM Glucose
	145 mM NaCl
Hypotonic buffer	5 mM KCl
79 mOsm, pH 7.2	1 mM $MgCl_2$
	2 mM $CaCl_2$
	10 mM HEPES-NaOH
	10 mM Glucose
	20 mM NaCl
Isotonic buffer	5 mM KCl
329 mOsm, pH 7.2	1 mM $MgCl_2$
	2 mM $CaCl_2$
	10 mM HEPES-NaOH
	10 mM Glucose
	20 mM NaCl
	250 mM Mannitol

Calcein fluorescence was normalized to fluorescence at the time point of stimulation and plotted over time. Fluorescence decrease was determined from the linear slope m ($\Delta F/\Delta t$) over 20 sec after maximal fluorescence increase. Rate of calcein fluorescence decrease was used as a measure of regulatory volume decrease.

In order to quantify the effect of chloride channel inhibitors or knock-out of LRRC8A on RVD, RVD was normalized accordingly to VRAC activity (see 7.3.1). Arctan function (α=arctan(m)) was used to convert slope m into the gradient angle α. The gradient angle of isotonic control (α_{iso}) was subtracted ($\alpha_{correct}=\alpha-\alpha_{iso}$) to correct for background activity. Then corrected gradient angles ($\alpha_{correct}$) were related to hypotonic control (α_{hypo}) and finally displayed in percent of hypotonic control (($\alpha_{correct}/\alpha_{hypotonic}$)*100), which was defined as normalized RVD (%).

7.3.3 Measuring changes in intracellular Ca^{2+} concentration using Fluo4-AM

To measure changes in intracellular Ca^{2+} concentration, 30 000 HaCaT cells were seeded in 96-well black-walled, clear bottom microplates and incubated at 37 °C and 5 % CO_2 for 24 h. Prior to measurement cells were incubated for 1 h with 100 µl culture medium containing 2 µM Ca^{2+}-sensitive dye Fluo4-AM and 250 µM sulfinpyrazone, to trap cationic Fluo4 inside the cell. Medium was removed and cells were washed once with 70 µl isotonic buffer (see Table 7-3) containing 250 µM sulfinpyrazone. Cells were covered with 50 µl isotonic buffer containing 250 µM sulfinpyrazone and plate was transferred into an automated fluorescence plate reader (FLIPR Tetra, Molecular Devices) and incubated for 15 min at 37 °C. Fluo4 fluorescence (excitation 470 – 495 nm, emission 515 – 575 nm) was continuously recorded every sec. After baseline recording for 20 sec, cells were stimulated by addition of 125 µl isotonic or hypotonic buffer (see Table 7-3). Addition of hypotonic buffer resulted in a decrease of extracellular osmolarity to 150 mOsm. For experiments without extracellular Ca^{2+}, 2 mM $CaCl_2$ were replaced by 2 mM $MgCl_2$ and additional 1 mM EGTA.

Intracellular Fluo4 fluorescence was normalized to fluorescence at the time point of stimulation and plotted over time. Maximal increase of fluorescence ($\Delta F=F_{max}-F_0$) from baseline fluorescence before stimulation (F_0) to maximal fluorescence after buffer addition (F_{max}) was normalized to baseline fluorescence ($\Delta F/F_0$), which was used as a relative measure of maximal increase of intracellular Ca^{2+} concentration.

Increase of intracellular Ca^{2+} concentration was normalized to assess the effect of extracellular Ca^{2+}. $\Delta F/F_0$ of isotonic control ($\Delta F/F_{0,iso}$) was subtracted from $\Delta F/F_0$ resulting from hypotonic stimulation ($\Delta F/F_{0,correct}=\Delta F/F_0-\Delta F/F_{0,iso}$) to correct for background activity. Then corrected values ($\Delta F/F_{0correct}$) were related to hypotonic control containing 2 mM $CaCl_2$ ($\Delta F/F_{0,hypo}$) and finally displayed in percent of hypotonic control ((($\Delta F/F_{0,correct})/(\Delta F/F_{0,hypo}$))*100), which was defined as normalized intracellular Ca^{2+} concentration (%).

7.3.4   Determination of the population doubling time

To acquire a growth curve for HaCaT keratinocytes, $0.6*10^6$ HaCaT-WT or HaCaT-*LRRC8A*$^{-/-}$ cells were seeded in five 6-well plates and cultured at 37 °C and 5 % CO_2 for up to five days. Every 24 h cells of one 6-well plate were thoroughly detached from the plate by incubation with TrypLE and collected in medium. Cell density was determined by automated cell counting (see 7.1.4) and total number of cells in the collected cell suspension was calculated by multiplying the cell density with the total volume of cell suspension. Cell number was plotted over cultivation time. The population doubling time t_d was calculated by $t_d=\ln(2)/\lambda$ with growth rate $\lambda=\ln(N_t/N_0)/t$ where N is cell number at the beginning (N_0) or at the time point t (N_t).

7.3.5   Determination of metabolic activity using WST-1

To access the metabolic activity of HaCaT cells, 800 HaCaT-WT or HaCaT-*LRRC8A*$^{-/-}$ cells were seeded in 96-well plates and cultured at 37 °C and 5 % CO_2 for 120 h. Water soluble tetrazolium 1 (WST-1) dye (Roche) was diluted 1:10 in culture medium, added to the cells and incubated for 2 h in the cell culture incubator. Metabolic active cells enzymatically convert WST-1 to water soluble formazan. Absorption of formazan was detected at 450 nm with a reference measurement at 650 nm. For normalization, absorption at 650 nm (A_{650}) was subtracted from absorption at 450 nm (A_{450}). To correct for background, absorption of reagents (DMEM and WST-1) was subtracted from wavelength-corrected absorption. Finally, corrected values were related to WT as control and displayed in percent.

7.3.6   Determination of DNA synthesis using BrdU incorporation

To determine DNA synthesis, 800 HaCaT-WT or HaCaT-*LRRC8A*$^{-/-}$ cells were seeded in 96-well plates and cultured at 37 °C and 5 % CO_2 for 96 h. DNA synthesis was determined by incorporation and immunological detection of 5-bromo-2'-deoxyuridine (BrdU) (Roche). 100 mM BrdU was added to the culture medium in the microplate and cells were cultured for additional 24 h. After incubation, medium was removed and cells were fixed and denatured with FixDenat solution for 30 min at RT. Peroxidase-coupled Anti-BrdU antibody was diluted 1:10 in antibody-dilution-solution, added to cells and incubated for 90 min at RT. Antibody solution was removed and cells were washed with washing buffer. Tetramethylbenzidine (TMB) substrate solution was added and incubated for 30 min at RT. TMB is oxidized by peroxidase to TMB diimine, whose absorption was measured at 370 nm and 490 nm as reference. Absorption at 490 nm (A_{490}) was subtracted from absorption at 370 nm (A_{370}). To correct for background, absorption of TMB substrate solution was subtracted from wavelength-corrected absorption. Finally, corrected values were related to WT as control and displayed in percent.

7.3.7 Determination of the outside-in barrier of epidermis equivalents

Epidermis equivalents were grown from HaCaT-WT and HaCaT-$LRRC8A^{-/-}$ cells as described in 7.1.11. To destroy the uppermost cell layer, epidermis equivalent were incubated for 1 h with 0.5 % (v/v) SDS solution 20 % (w/v) that was then thoroughly removed. Each well of a 24-well plate was filled with 1 ml PBS and transwell plates with epidermis equivalents were placed in the wells with contact to PBS. SDS-treated and untreated epidermis equivalents were covered with 125 µl 1 M Lucifer yellow solution and incubated in a cell culture incubator for 3 h. Every 60 min three times 40 µl PBS were removed from the lower cavity and transferred into three wells of a 96-well black-walled, clear bottom microplate. PBS was refilled with 120 µl PBS. In addition, Lucifer yellow solution was used to obtain a standard curve in a concentration range from 0.38 – 0.02 mM. 96-well plate was stored in the dark until measurement in the fluorescence plate reader Flex Station 2 (Molecular Devices) with excitation at 425 nm and emission at 538 nm. Concentration of Lucifer yellow was calculated using the linear range of the standard curve and plotted over time.

After 3 h of incubation, residual Lucifer yellow solution was removed from the epidermis equivalents. Epidermis equivalents were washed with PBS, fixed in 4.5 % (v/v) PFA and stored in the dark until embedded in paraffin for histochemical procedure (see 7.2.10). Lucifer yellow treated epidermis equivalents were not used for immunohistochemical detection of different proteins, but only counterstained with DAPI for detection of nuclei.

7.3.8 Statistical analysis

Statistical significance was analyzed by two-sided, unpaired Student's t-tests without correction (Figure 5-17 B and Figure 5-19) or with Welch's correction (Figure 5-17 A) when two groups were compared. When more than two groups were compared, which is the case for all other measurements, one-way ANOVA with Bonferroni's correction was used to determine statistical significance. Statistical significance was indicated as ns – not significant, * $P<0.05$, ** $P<0.01$ and *** $P<0.001$.

7.4 Material

Table 7-4: Chemicals and reagents

Substance	Supplier
Acrylamide/bisacrylamide (37.5:1) 4K solution	AppliChem
Acetic acid	AppliChem
Agarose	Biozym Scientific
Amersham Protran Western Blot Nitrocellulose membrane	GE Healthcare
Ammonium persulfate (APS)	Sigma-Aldrich Chemie
β-Mercaptoethanol	AppliChem
Bromophenol blue	Sigma-Aldrich Chemie
$CaCl_2$	AppliChem
Calcein-AM	Thermo Fisher Scientific
Carbeneoxolone (CBX)	Tocris Bioscience
Citrate buffer	Zytomed Systems
DAPI	Sigma-Aldrich Chemie
DCPIB	Tocris Bioscience
Deoxynucleotide (dNTP) Solution Mix	New England Biolabs
DIDS	Tocris Bioscience
Dithiothreitol (DTT)	AppliChem
DNA ladder 100 bp	New England Biolabs
EDTA $2xH_2O$	AppliChem
EDTA solution	Zytomed Systems
EGTA	AppliChem
Enthelan	Merck Millipore
Ethanol, abs. molecular biology grade	AppliChem
Fluo-4-AM	Thermo Fisher Scientific
Fluoromount-G	eBioscience
Glucose	AppliChem
Glycerol	AppliChem
Glycine	AppliChem
Hematoxylin	Sigma-Aldrich Chemie
HEPES	AppliChem
Histofine Simple Stain AP Multi	Medac Diagnostika
Isopropanol	AppliChem
KCl	AppliChem
Mannitol	AppliChem
Methanol	AppliChem
$MgCl_2$	AppliChem
Midori Green	Biozym Scientific
NaCl	AppliChem
NaI	Sigma-Aldrich Chemie
Niflumic acid (NFA)	Sigma-Aldrich Chemie
Nonidet P-40	AppliChem
Orange G (6x)	AppliChem
peqGOLD Protein Marker IV	VWR Chemicals

Phusion HF DNA Polymerase	New England Biolabs
Ponceaus S	Sigma-Aldrich Chemie
Protease inhibitor mix M	Serva Electrophoresis
RNAse-free water	Qiagen
Roti-Histofix 4,5 % (v/v), acid free (ph 7) - phosphate buffered formaldehyd	Carl Roth GmbH
Signal stain buffer	Cell Signaling Technology
Skim milk	Merck
Sodium deoxycholate	Sigma-Aldrich Chemie
Sodium dodecyl sulfate (SDS) solution, 20 % (w/v)	AppliChem
Sulfinpyrazone	AppliChem
TBS buffer	Zytomed Systems
TEMED	Thermo Fisher Scientific
Tris-Base	Sigma-Aldrich Chemie
Tris-Glycine SDS running buffer	Thermo Fisher Scientific
Triton X-100	AppliChem
Tween20	Sigma-Aldrich Chemie
Xylol	Sigma-Aldrich Chemie

Table 7-5: Chemicals and reagents used for cell culture

Substance	Supplier
CASY ton solution	OLS OMNI Life Science
CnT prime 3D medium	CELLnTEC Advanced Cell Systems
CnT prime epithelial culture medium	CELLnTEC Advanced Cell Systems
Dimethyl sulfoxide (DMSO)	AppliChem
Dulbecco`s Modified Eagle Medium (DMEM) high glucose 4.5 g/L	PAA Laboratories
Dulbecco`s Modified Eagle Medium (DMEM) high glucose 4.5 g/L, 0 mM $CaCl_2$	PAA Laboratories
Fetal bovine serum superior (FCS)	Biochrom
HEPES BSS	PromoCell
Keratinocyte Growth Medium 2	PromoCell
L-Glutamine	PAA Laboratories
Phosphate buffered saline (PBS)	Lonza
Polybrene	Merck Millipore
Puromycin	Thermo Fisher Scientific
Supplement Mix	PromoCell
Trybane blue	Sigma-Aldrich Chemie
TrypLE express enzyme	Thermo Fisher Scientific
Trypsin Neutralization Solution	PromoCell
Trypsin/EDTA	PromoCell

Table 7-6: Kits and enzymes

Kit/Enzyme	Supplier
Cell Proliferation ELISA, BrdU	Merck - Roche
Cell Proliferation Reagent WST-1	Merck - Roche
ECL reagent: Immobilon Western Chemiluminescent HRP Substrate	Merck Millipore
ECL reagent: SuperSignal West Femto Maximum Sensitivity Substrate	Thermo Fisher Scientific
NucleoSpin RNA II	Macherey-Nagel
Phusion HF DNA Polymerase	New England Biolabs
Pierce BCA Protein Assay Kit	Thermo Fisher Scientific
Protoscript M-MuLV First Strand cDNA Synthesis	New England Biolabs
QiaQuick PCR Purification	Qiagen
QuickExtract DNA Extraction Solution	Epicentre
Substrate solution permanent AP red kit	Zytomed Systems

Table 7-7: Oligonucleotide primer for target site-specific PCR and DNA sequencing

Gene	Forward 5`- 3`	Reverse 5`- 3`	Size
LRRC8A	TGGTTTCCCAGCCAAGTG	GCGGGAATTTGAACCAGAAG	657 bp

Table 7-8: Single guide RNA targeting LRRC8A

Name	Target sequence
LRRC8A#1	GCTGCGTGTCCGCAAAGTAG
LRRC8A#2	CCGGCACCAGTACAACTACG

Table 7-9: Adenoviruses

Generated by Sirion Biotech

Name	Genes
Ad5-CMV-Cas9-wt-2A-OFP	Cas9 nuclease and orange fluorescent protein
Ad5-U6-sgRNA-LRRC8A#1-U6-sgRNA-LRRC8A#2	sgRNA targeting LRRC8A in two positions
Ad5-CMV-hsYFP	halide-sensitive yellow fluorescent protein

Table 7-10: Antibodies used in Western blot and immunohistochemistry

Antigen	Supplier	Product No	Dilution	Species	Application
α-Tubulin	Sigma Aldrich Corporation	T9026	1:500	mouse	Western Blot
β-Actin	Sigma Aldrich Corporation	A1978	1:10 000	mouse	Western Blot
Keratin 10	abcam	ab76318	1:10 000 (WB) 1:5 000 (IHC)	rabbit	Western Blot, Immunohisto-chemistry
Involucrin	abcam	ab20202 (SY8)	1:10 000	mouse	Western Blot
Involucrin	Santa Cruz Biotechnology	sc-28557	1:3 000	rabbit	Immunohisto-chemistry
LRRC8A	Novus Biologicals	NBP2-32158	1:500 (WB) 1:100 (IHC)	rabbit	Western Blot, Immunohisto-chemistry
RPLP0	abcam	ab192866	1:100	rabbit	Western Blot
YWHAZ	Santa Cruz Biotechnology	sc-293415	1:100	mouse	Western Blot

Table 7-11: Equipment

Name	Model	Manufacturer
Cell counter	CASY Model TT	OLS OMNI Life Science
Fluorescence lamp	U-RFL-T	Olympus
Fluorescence microplate reader	Flex Station 2 and FLIPR tetra	Molecular Devices
Fluorescence microscope	IX71S8F	Olympus
Gel documentation system	Infinity Serial N	Vilber
Inverted contrasting microscope	DM IL	Leica microsystems
Microscope camera	XC50	Olympus
PCR Thermo cycler	Mastercycler gradient	Eppendorf
Spectrophotometer	NanoDrop-1060 Spectrometer	PeqLab
Western blot documentation system	Fusion Fx7	Vilber

8   Supplementary Information

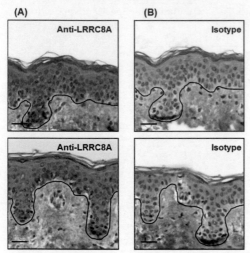

Supplement Figure S1: Immunohistological staining of LRRC8A in human skin biopsies

Punch biopsies from two additional healthy donors were taken and sections were subjected to immunohistological staining with Anti-LRRC8A antibody (A) or isotype control antibody (B) and nuclei staining with hematoxylin (blue). LRRC8A antibody binding is visualized by red staining and shows preferential expression of *LRRC8A* in the lowest epidermal layer, the *stratum basale*. Black line illustrates the border between dermis and epidermis. Bars represent 30 μm.

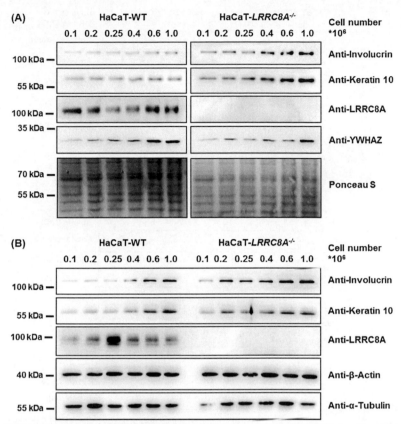

Supplement Figure S2: Differentiation of HaCaT-WT and HaCaT-LRRC8A-/- cells upon post-confluent grwoth

HaCaT-WT and HaCaT-*LRRC8A^-/-* cells were seeded in increasing cell densities from $0.1*10^6$ – $1.0*10^6$ cells/cavity. Whole cell protein lysates were isolated and subjected to Western blot analysis. The differentiation markers involucrin and keratin 10 increased with increasing cell density indicating differentiation of HaCaT cells. Earlier appearance and stronger intensity of involucrin and keratin 10 in HaCaT-*LRRC8A^-/-* cells points to earlier onset of differentiation in the absence of LRRC8A, which was confirmed by detection of LRRC8A. Western blot analysis of YWHAZ, β-actin or α-tubulin were used as loading control as well as membrane staining using Ponceau S. (A) and (B) show Western blots of independently grown and analyzed cells.

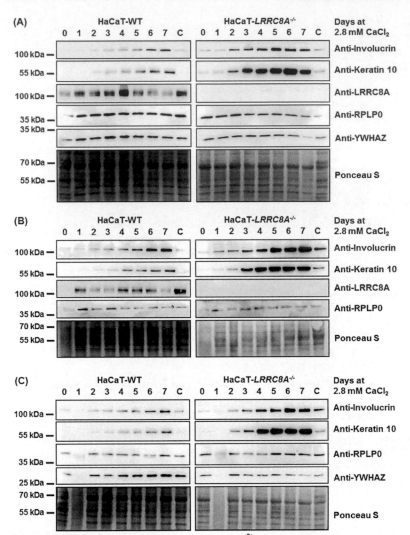

Supplement Figure S3: Western blot analysis of Ca²⁺-induced differentiation of HaCaT-WT and HaCaT-LRRC8A⁻/⁻ cells

HaCaT-WT and HaCaT-*LRRC8A*⁻/⁻ cells were seeded in medium containing 0.03 mM CaCl₂ before changing medium to 2.8 mM CaCl₂ the day after seeding. As control, cells were grown in medium containing 0.03 mM CaCl₂ for seven days (marked as C). Whole cell protein lysates were subjected to Western blot. Differentiation markers involucrin and keratin 10 increase over cultivation time indicating differentiation of HaCaT cells. Increase of both markers started earlier and increased stronger in HaCaT-*LRRC8A*⁻/⁻ compared to HaCaT-WT cells indicating involvement of LRRC8A in HaCaT keratinocyte differentiation. Absence of LRRC8A in HaCaT-*LRRC8A*⁻/⁻ cells was confirmed. Western blot analysis of RPLP0 and YWHAZ as well as membrane staining with Ponceau S was used as loading control. (A),(B) and (C) show Western blots of independently grown and analyzed cells.

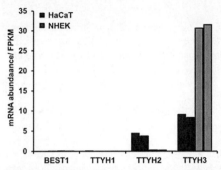

Supplement Figure S4: Expression of BEST1 and TTYH1-3 in human keratinocytes

Total mRNA isolated from HaCaT and NHEK cells was used for RNA sequencing to determine mRNA transcript abundances of bestrophin 1 (BEST1) and *Drosophila melanogaster* tweety homologue 1-3 (TTYH 1-3). Transcript abundances are displayed in fragments per kilobase of exon per million fragments mapped (FPKM) from two independent RNA preparations per cell line.

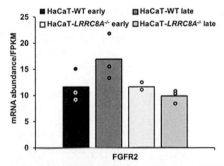

Supplement Figure S5: Expression of FGFR2 during differentiation of HaCaT-WT and HaCaT-LRRC8A$^{-/-}$ cells

HaCaT-WT and HaCaT-*LRRC8A$^{-/-}$* cells were differentiated by post-confluent growth and RNA of earliest and latest differentiation state was subjected to RNA sequencing. mRNA abundances of keratinocyte growth factor receptor (FGFR2) is displayed in fragments per kilobase of exon per million fragments mapped (FPKM) as the mean (bars) of the single values (dots) of three independent RNA preparations.

9   References

1.   Ng, K.W. and Lau, W.M., *Skin Deep: The Basics of Human Skin Structure and Drug Penetration*, in *Percutaneous Penetration Enhancers Chemical Methods in Penetration Enhancement: Drug Manipulation Strategies and Vehicle Effects*, N. Dragicevic and H.I. Maibach, Editors. 2015, Springer Berlin Heidelberg: Berlin, Heidelberg. p. 3-11.

2.   Freinkel, R.K. and Woodley, D., *The biology of the skin*. 2001, New York: Parthenon Pub. Group. 432 p.

3.   Baroni, A., Buommino, E., De Gregorio, V., Ruocco, E., Ruocco, V., and Wolf, R., *Structure and function of the epidermis related to barrier properties*. Clinics in Dermatology, 2012. **30**(3): p. 257-262.

4.   Moll, I., Roessler, M., Brandner, J.M., Eispert, A.-C., Houdek, P., and Moll, R., *Human Merkel cells – aspects of cell biology, distribution and functions*. European Journal of Cell Biology, 2005. **84**(2): p. 259-271.

5.   Gonzales, K.A.U. and Fuchs, E., *Skin and Its Regenerative Powers: An Alliance between Stem Cells and Their Niche*. Developmental cell, 2017. **43**(4): p. 387-401.

6.   Barrandon, Y. and Green, H., *Three clonal types of keratinocyte with different capacities for multiplication*. Proceedings of the National Academy of Sciences of the United States of America, 1987. **84**(8): p. 2302-2306.

7.   Blanpain, C. and Fuchs, E., *Epidermal homeostasis: a balancing act of stem cells in the skin*. Nature reviews. Molecular cell biology, 2009. **10**(3): p. 207-217.

8.   Miroshnikova, Y.A., Le, H.Q., Schneider, D., Thalheim, T., Rübsam, M., Bremicker, N., Polleux, J., Kamprad, N., Tarantola, M., Wang, I., Balland, M., Niessen, C.M., Galle, J., and Wickström, S.A., *Adhesion forces and cortical tension couple cell proliferation and differentiation to drive epidermal stratification*. Nature Cell Biology, 2018. **20**(1): p. 69-80.

9.   Mesa, K.R., Kawaguchi, K., Cockburn, K., Gonzalez, D., Boucher, J., Xin, T., Klein, A.M., and Greco, V., *Homeostatic Epidermal Stem Cell Self-Renewal Is Driven by Local Differentiation*. Cell Stem Cell, 2018. **23**(5): p. 677-686.e4.

10.   Watt, F.M., *Epidermal stem cells: markers, patterning and the control of stem cell fate*. Philosophical transactions of the Royal Society of London. Series B, Biological sciences, 1998. **353**(1370): p. 831-837.

11.   Hsu, Y.-C., Li, L., and Fuchs, E., *Emerging interactions between skin stem cells and their niches*. Nature Medicine, 2014. **20**: p. 847.

12.   Kolly, C., Suter, M.M., and Müller, E.J., *Proliferation, Cell Cycle Exit, and Onset of Terminal Differentiation in Cultured Keratinocytes: Pre-Programmed Pathways in Control of C-Myc and Notch1 Prevail Over Extracellular Calcium Signals*. Journal of Investigative Dermatology, 2005. **124**(5): p. 1014-1025.

13.   Rangarajan, A., Talora, C., Okuyama, R., Nicolas, M., Mammucari, C., Oh, H., Aster, J.C., Krishna, S., Metzger, D., Chambon, P., Miele, L., Aguet, M., Radtke, F., and Dotto, G.P., *Notch signaling is a direct determinant of keratinocyte growth arrest and entry into differentiation*. The EMBO Journal, 2001. **20**(13): p. 3427-3436.

14.   Calautti, E., Li, J., Saoncella, S., Brissette, J.L., and Goetinck, P.F., *Phosphoinositide 3-Kinase Signaling to Akt Promotes Keratinocyte Differentiation Versus Death*. Journal of Biological Chemistry, 2005. **280**(38): p. 32856-32865.

15.   Buerger, C., Shirsath, N., Lang, V., Berard, A., Diehl, S., Kaufmann, R., Boehncke, W.-H., and Wolf, P., *Inflammation dependent mTORC1 signaling interferes with the switch from keratinocyte proliferation to differentiation*. PLOS ONE, 2017. **12**(7): p. e0180853.

16. Dazard, J.-E., Piette, J., Basset-Seguin, N., Blanchard, J.-M., and Gandarillas, A., *Switch from p53 to MDM2 as differentiating human keratinocytes lose their proliferative potential and increase in cellular size.* Oncogene, 2000. **19**(33): p. 3693-3705.

17. Zanet, J., Freije, A., Ruiz, M., Coulon, V., Sanz, J.R., Chiesa, J., and Gandarillas, A., *A Mitosis Block Links Active Cell Cycle with Human Epidermal Differentiation and Results in Endoreplication.* PLOS ONE, 2010. **5**(12): p. e15701.

18. Gandarillas, A., Davies, D., and Blanchard, J.-M., *Normal and c-Myc-promoted human keratinocyte differentiation both occur via a novel cell cycle involving cellular growth and endoreplication.* Oncogene, 2000. **19**(29): p. 3278-3289.

19. Bragulla, H.H. and Homberger, D.G., *Structure and functions of keratin proteins in simple, stratified, keratinized and cornified epithelia.* Journal of anatomy, 2009. **214**(4): p. 516-559.

20. Wang, F., Zieman, A., and Coulombe, P.A., *Skin Keratins.* Methods in enzymology, 2016. **568**: p. 303-350.

21. Kartasova, T., Roop, D.R., Holbrook, K.A., and Yuspa, S.H., *Mouse differentiation-specific keratins 1 and 10 require a preexisting keratin scaffold to form a filament network.* The Journal of cell biology, 1993. **120**(5): p. 1251-1261.

22. Eckhart, L., Lippens, S., Tschachler, E., and Declercq, W., *Cell death by cornification.* Biochimica et Biophysica Acta (BBA) - Molecular Cell Research, 2013. **1833**(12): p. 3471-3480.

23. Rinnerthaler, M., Streubel, M.K., Bischof, J., and Richter, K., *Skin aging, gene expression and calcium.* Experimental Gerontology, 2015. **68**: p. 59-65.

24. Broome, A.-M., Ryan, D., and Eckert, R.L., *S100 protein subcellular localization during epidermal differentiation and psoriasis.* The journal of histochemistry and cytochemistry : official journal of the Histochemistry Society, 2003. **51**(5): p. 675-685.

25. Nithya, S., Radhika, T., and Jeddy, N., *Loricrin - an overview.* Journal of oral and maxillofacial pathology : JOMFP, 2015. **19**(1): p. 64-68.

26. Candi, E., Schmidt, R., and Melino, G., *The cornified envelope: a model of cell death in the skin.* Nature Reviews Molecular Cell Biology, 2005. **6**(4): p. 328-340.

27. Manabe, M. and O'Guin, W.M., *Keratohyalin, Trichohyalin and Keratohyalin-Trichohyalin Hybrid Granules: An Overview.* The Journal of Dermatology, 1992. **19**(11): p. 749-755.

28. Dale, B.A., Resing, K.A., and Lonsdale-Eccles, J.D., *Filaggrin: A Keratin Filament Associated Proteina.* Annals of the New York Academy of Sciences, 1985. **455**(1): p. 330-342.

29. Elias, P.M., Gruber, R., Crumrine, D., Menon, G., Williams, M.L., Wakefield, J.S., Holleran, W.M., and Uchida, Y., *Formation and functions of the corneocyte lipid envelope (CLE).* Biochimica et biophysica acta, 2014. **1841**(3): p. 314-318.

30. Sanford, J.A. and Gallo, R.L., *Functions of the skin microbiota in health and disease.* Seminars in immunology, 2013. **25**(5): p. 370-377.

31. Nestle, F.O., Di Meglio, P., Qin, J.-Z., and Nickoloff, B.J., *Skin immune sentinels in health and disease.* Nature reviews. Immunology, 2009. **9**(10): p. 679-691.

32. Proksch, E., Brandner, J.M., and Jensen, J.-M., *The skin: an indispensable barrier.* Experimental Dermatology, 2008. **17**(12): p. 1063-1072.

33. Niessen, C.M., *Tight Junctions/Adherens Junctions: Basic Structure and Function.* Journal of Investigative Dermatology, 2007. **127**(11): p. 2525-2532.

34.   Brandner, J.M., Kief, S., Wladykowski, E., Houdek, P., and Moll, I., *Tight Junction Proteins in the Skin.* Skin Pharmacology and Physiology, 2006. **19**(2): p. 71-77.

35.   Yokouchi, M., Atsugi, T., Logtestijn, M.v., Tanaka, R.J., Kajimura, M., Suematsu, M., Furuse, M., Amagai, M., and Kubo, A., *Epidermal cell turnover across tight junctions based on Kelvin's tetrakaidecahedron cell shape.* eLife, 2016. **5**: p. e19593.

36.   Chiba, T., Nakahara, T., Kohda, F., Ichiki, T., Manabe, M., and Furue, M., *Measurement of trihydroxy-linoleic acids in stratum corneum by tape-stripping: Possible biomarker of barrier function in atopic dermatitis.* PloS one, 2019. **14**(1): p. e0210013-e0210013.

37.   Pierard, G.E., Goffin, V., Hermanns-Le, T., and Pierard-Franchimont, C., *Corneocyte desquamation.* Int J Mol Med, 2000. **6**(2): p. 217-21.

38.   Boehncke, W.-H. and Schön, M.P., *Psoriasis.* The Lancet, 2015. **386**(9997): p. 983-994.

39.   Deng, Y., Chang, C., and Lu, Q., *The Inflammatory Response in Psoriasis: a Comprehensive Review.* Clin Rev Allergy Immunol, 2016. **50**(3): p. 377-89.

40.   Caspers, P.J., Lucassen, G.W., and Puppels, G.J., *Combined In Vivo Confocal Raman Spectroscopy and Confocal Microscopy of Human Skin.* Biophysical Journal, 2003. **85**(1): p. 572-580.

41.   El-Chami, C., Haslam, I.S., Steward, M.C., and O'Neill, C.A., *Role of organic osmolytes in water homoeostasis in skin.* Experimental Dermatology, 2014. **23**(8): p. 534-537.

42.   Verdier-Sévrain, S. and Bonté, F., *Skin hydration: a review on its molecular mechanisms.* Journal of Cosmetic Dermatology, 2007. **6**(2): p. 75-82.

43.   Kippenberger, S., Loitsch, S., Guschel, M., Müller, J., Kaufmann, R., and Bernd, A., *Hypotonic stress induces E-cadherin expression in cultured human keratinocytes.* FEBS Letters, 2005. **579**(1): p. 207-214.

44.   Häussinger, D., *The role of cellular hydration in the regulation of cell function.* Biochemical Journal, 1996. **313**(3): p. 697-710.

45.   Lang, F., Busch, G.L., Ritter, M., Völkl, H., Waldegger, S., Gulbins, E., and Häussinger, D., *Functional Significance of Cell Volume Regulatory Mechanisms.* Physiological Reviews, 1998. **78**(1): p. 247-306.

46.   Dubois, J.M. and Rouzaire-Dubois, B., *Roles of cell volume in molecular and cellular biology.* Prog Biophys Mol Biol, 2012. **108**(3): p. 93-7.

47.   Verkman, A.S., *Aquaporins at a glance.* Journal of Cell Science, 2011. **124**(13): p. 2107-2112.

48.   Lang, F., Busch, G.L., and Völkl, H., *The Diversity of Volume Regulatory Mechanisms.* Cellular Physiology and Biochemistry, 1998. **8**(1-2): p. 1-45.

49.   Sardini, A., Amey, J.S., Weylandt, K.H., Nobles, M., Valverde, M.A., and Higgins, C.F., *Cell volume regulation and swelling-activated chloride channels.* Biochim Biophys Acta, 2003. **1618**(2): p. 153-62.

50.   Jentsch, T.J., *VRACs and other ion channels and transporters in the regulation of cell volume and beyond.* Nature Reviews Molecular Cell Biology, 2016. **17**: p. 293.

51.   Strange, K., *Cellular volume homeostasis.* Adv Physiol Educ, 2004. **28**(1-4): p. 155-9.

52.   Strange, K., *Cellular and molecular physiology of cell volume regulation.* 1994, Boca Raton: CRC Press. 400 p.

53.   Lutter, D., Ullrich, F., Lueck, J.C., Kempa, S., and Jentsch, T.J., *Selective transport of neurotransmitters and modulators by distinct volume-regulated LRRC8 anion channels.* Journal of Cell Science, 2017. **130**(6): p. 1122-1133.

54. Hoffmann, E.K., Lambert, I.H., and Pedersen, S.F., *Physiology of cell volume regulation in vertebrates*. Physiol Rev, 2009. **89**(1): p. 193-277.

55. Janeke, G., Siefken, W., Carstensen, S., Springmann, G., Bleck, O., Steinhart, H., Hoger, P., Wittern, K.P., Wenck, H., Stab, F., Sauermann, G., Schreiner, V., and Doering, T., *Role of taurine accumulation in keratinocyte hydration*. J Invest Dermatol, 2003. **121**(2): p. 354-61.

56. Stutzin, A. and Hoffmann, E., *Swelling-activated ion channels: Functional regulation in cell-swelling, proliferation and apoptosis*. Acta physiologica (Oxford, England), 2006. **187**: p. 27-42.

57. Jakab, M., Fuerst, J., Gschwentner, M., Bottà, G., Garavaglia, M.L., Bazzini, C., Rodighiero, S., Meyer, G., Eichmueller, S., Woell, E., Chwatal, S., Ritter, M., and Paulmichl, M., *Mechanisms Sensing and Modulating Signals Arising From Cell Swelling*. Cellular Physiology and Biochemistry, 2002. **12**(5-6): p. 235-258.

58. McCarty, N.A. and O'Neil, R.G., *Calcium signaling in cell volume regulation*. Physiol Rev, 1992. **72**(4): p. 1037-61.

59. Azorin, N., Raoux, M., Rodat-Despoix, L., Merrot, T., Delmas, P., and Crest, M., *ATP signalling is crucial for the response of human keratinocytes to mechanical stimulation by hypo-osmotic shock*. Experimental Dermatology, 2011. **20**(5): p. 401-407.

60. Gönczi, M., Szentandrássy, N., Fülöp, L., Telek, A., Szigeti, G.P., Magyar, J., Bíró, T., Nánási, P.P., and Csernoch, L., *Hypotonic stress influence the membrane potential and alter the proliferation of keratinocytes in vitro*. Experimental Dermatology, 2007. **16**(4): p. 302-310.

61. Zholos, A., Beck, B., Sydorenko, V., Lemonnier, L., Bordat, P., Prevarskaya, N., and Skryma, R., *Ca2+- and Volume-sensitive Chloride Currents Are Differentially Regulated by Agonists and Store-operated Ca2+ Entry*. The Journal of General Physiology, 2005. **125**(2): p. 197-211.

62. Habela, C.W. and Sontheimer, H., *Cytoplasmic Volume Condensation Is an Integral Part of Mitosis*. Cell Cycle, 2007. **6**(13): p. 1613-1620.

63. Stewart, M.P., Helenius, J., Toyoda, Y., Ramanathan, S.P., Muller, D.J., and Hyman, A.A., *Hydrostatic pressure and the actomyosin cortex drive mitotic cell rounding*. Nature, 2011. **469**: p. 226.

64. Cadart, C., Zlotek-Zlotkiewicz, E., Le Berre, M., Piel, M., and Matthews, Helen K., *Exploring the Function of Cell Shape and Size during Mitosis*. Developmental Cell, 2014. **29**(2): p. 159-169.

65. Pendergrass, W.R., Angello, J.C., Kirschner, M.D., and Norwood, T.H., *The relationship between the rate of entry into S phase, concentration of DNA polymerase α, and cell volume in human diploid fibroblast-like monokaryon cells*. Experimental Cell Research, 1991. **192**(2): p. 418-425.

66. Putney, L.K. and Barber, D.L., *Na-H Exchange-dependent Increase in Intracellular pH Times G2/M Entry and Transition*. Journal of Biological Chemistry, 2003. **278**(45): p. 44645-44649.

67. Takahashi, A., Yamaguchi, H., and Miyamoto, H., *Change in K+ current of HeLa cells with progression of the cell cycle studied by patch-clamp technique*. American Journal of Physiology-Cell Physiology, 1993. **265**(2): p. C328-C336.

68. Chen, L., Wang, L., Zhu, L., Nie, S., Zhang, J., Zhong, P., Cai, B., Luo, H., and Jacob, T.J.C., *Cell cycle-dependent expression of volume-activated chloride currents in*

nasopharyngeal carcinoma cells. American Journal of Physiology-Cell Physiology, 2002. **283**(4): p. C1313-C1323.

69. Jackson, P.S. and Strange, K., *Characterization of the voltage-dependent properties of a volume-sensitive anion conductance.* J Gen Physiol, 1995. **105**(5): p. 661-76.

70. Bond, T., Basavappa, S., Christensen, M., and Strange, K., *ATP dependence of the ICl,swell channel varies with rate of cell swelling. Evidence for two modes of channel activation.* J Gen Physiol, 1999. **113**(3): p. 441-56.

71. Kimelberg, H.K., Goderie, S.K., Higman, S., Pang, S., and Waniewski, R.A., *Swelling-induced release of glutamate, aspartate, and taurine from astrocyte cultures.* J Neurosci, 1990. **10**(5): p. 1583-91.

72. Chen, L., Konig, B., Liu, T., Pervaiz, S., Razzaque, Y.S., and Stauber, T., *More than just a pressure release valve: physiological roles of volume-regulated LRRC8 anion channels.* Biol Chem, 2019.

73. Strange, K., Yamada, T., and Denton, J.S., *A 30-year journey from volume-regulated anion currents to molecular structure of the LRRC8 channel.* J Gen Physiol, 2019. **151**(2): p. 100-117.

74. Akita, T., Fedorovich, S.V., and Okada, Y., *Ca2+ nanodomain-mediated component of swelling-induced volume-sensitive outwardly rectifying anion current triggered by autocrine action of ATP in mouse astrocytes.* Cell Physiol Biochem, 2011. **28**(6): p. 1181-90.

75. Nilius, B. and Droogmans, G., *Amazing chloride channels: an overview.* Acta Physiol Scand, 2003. **177**(2): p. 119-47.

76. Pedersen, S.F., Klausen, T.K., and Nilius, B., *The identification of a volume-regulated anion channel: an amazing Odyssey.* Acta Physiol (Oxf), 2015. **213**(4): p. 868-81.

77. Kunzelmann, K., *TMEM16, LRRC8A, bestrophin: chloride channels controlled by Ca(2+) and cell volume.* Trends Biochem Sci, 2015. **40**(9): p. 535-43.

78. Bae, Y., Kim, A., Cho, C.-H., Kim, D., Jung, H.-G., Kim, S.-S., Yoo, J., Park, J.-Y., and Hwang, E.M., *TTYH1 and TTYH2 Serve as LRRC8A-Independent Volume-Regulated Anion Channels in Cancer Cells.* Cells, 2019. **8**(6): p. 562.

79. Han, Y.-E., Kwon, J., Won, J., An, H., Jang, M.W., Woo, J., Lee, J.S., Park, M.G., Yoon, B.-E., Lee, S.E., Hwang, E.M., Jung, J.-Y., Park, H., Oh, S.-J., and Lee, C.J., *Tweety-homolog (Ttyh) Family Encodes the Pore-forming Subunits of the Swelling-dependent Volume-regulated Anion Channel (VRACswell) in the Brain.* Exp Neurobiol, 2019. **28**(2): p. 183-215.

80. Stotz, S.C. and Clapham, D.E., *Anion-sensitive fluorophore identifies the Drosophila swell-activated chloride channel in a genome-wide RNA interference screen.* PLoS One, 2012. **7**(10): p. e46865.

81. Milenkovic, A., Brandl, C., Milenkovic, V.M., Jendryke, T., Sirianant, L., Wanitchakool, P., Zimmermann, S., Reiff, C.M., Horling, F., Schrewe, H., Schreiber, R., Kunzelmann, K., Wetzel, C.H., and Weber, B.H., *Bestrophin 1 is indispensable for volume regulation in human retinal pigment epithelium cells.* Proc Natl Acad Sci U S A, 2015. **112**(20): p. E2630-9.

82. Chien, L.T. and Hartzell, H.C., *Rescue of volume-regulated anion current by bestrophin mutants with altered charge selectivity.* J Gen Physiol, 2008. **132**(5): p. 537-46.

83. Hartzell, H.C., Qu, Z., Yu, K., Xiao, Q., and Chien, L.T., *Molecular physiology of bestrophins: multifunctional membrane proteins linked to best disease and other retinopathies.* Physiol Rev, 2008. **88**(2): p. 639-72.

84. Caputo, A., Caci, E., Ferrera, L., Pedemonte, N., Barsanti, C., Sondo, E., Pfeffer, U., Ravazzolo, R., Zegarra-Moran, O., and Galietta, L.J., *TMEM16A, a membrane protein associated with calcium-dependent chloride channel activity.* Science, 2008. **322**(5901): p. 590-4.

85. Hartzell, H.C., Yu, K., Xiao, Q., Chien, L.T., and Qu, Z., *Anoctamin/TMEM16 family members are Ca2+-activated Cl- channels.* J Physiol, 2009. **587**(Pt 10): p. 2127-39.

86. Almaca, J., Tian, Y., Aldehni, F., Ousingsawat, J., Kongsuphol, P., Rock, J.R., Harfe, B.D., Schreiber, R., and Kunzelmann, K., *TMEM16 proteins produce volume-regulated chloride currents that are reduced in mice lacking TMEM16A.* J Biol Chem, 2009. **284**(42): p. 28571-8.

87. Suzuki, M. and Mizuno, A., *A novel human Cl(-) channel family related to Drosophila flightless locus.* J Biol Chem, 2004. **279**(21): p. 22461-8.

88. Qiu, Z., Dubin, Adrienne E., Mathur, J., Tu, B., Reddy, K., Miraglia, Loren J., Reinhardt, J., Orth, Anthony P., and Patapoutian, A., *SWELL1, a Plasma Membrane Protein, Is an Essential Component of Volume-Regulated Anion Channel.* Cell, 2014. **157**(2): p. 447-458.

89. Voss, F.K., Ullrich, F., Munch, J., Lazarow, K., Lutter, D., Mah, N., Andrade-Navarro, M.A., von Kries, J.P., Stauber, T., and Jentsch, T.J., *Identification of LRRC8 heteromers as an essential component of the volume-regulated anion channel VRAC.* Science, 2014. **344**(6184): p. 634-8.

90. Abascal, F. and Zardoya, R., *LRRC8 proteins share a common ancestor with pannexins, and may form hexameric channels involved in cell-cell communication.* Bioessays, 2012. **34**(7): p. 551-60.

91. Behe, P., Foote, J.R., Levine, A.P., Platt, C.D., Chou, J., Benavides, F., Geha, R.S., and Segal, A.W., *The LRRC8A Mediated "Swell Activated" Chloride Conductance Is Dispensable for Vacuolar Homeostasis in Neutrophils.* Frontiers in Pharmacology, 2017. **8**(262).

92. Planells-Cases, R., Lutter, D., Guyader, C., Gerhards, N.M., Ullrich, F., Elger, D.A., Kucukosmanoglu, A., Xu, G., Voss, F.K., Reincke, S.M., Stauber, T., Blomen, V.A., Vis, D.J., Wessels, L.F., Brummelkamp, T.R., Borst, P., Rottenberg, S., and Jentsch, T.J., *Subunit composition of VRAC channels determines substrate specificity and cellular resistance to Pt-based anti-cancer drugs.* The EMBO Journal, 2015. **34**(24): p. 2993-3008.

93. Bach, M.D., Sørensen, B.H., and Lambert, I.H., *Stress-induced modulation of volume-regulated anions channels in human alveolar carcinoma cells.* Physiological Reports, 2018. **6**(19): p. e13869.

94. Zhang, Y., Xie, L., Gunasekar, S.K., Tong, D., Mishra, A., Gibson, W.J., Wang, C., Fidler, T., Marthaler, B., Klingelhutz, A., Abel, E.D., Samuel, I., Smith, J.K., Cao, L., and Sah, R., *SWELL1 is a regulator of adipocyte size, insulin signalling and glucose homeostasis.* Nature Cell Biology, 2017. **19**: p. 504.

95. Kang, C., Xie, L., Gunasekar, S.K., Mishra, A., Zhang, Y., Pai, S., Gao, Y., Kumar, A., Norris, A.W., Stephens, S.B., and Sah, R., *SWELL1 is a glucose sensor regulating β-cell excitability and systemic glycaemia.* Nature Communications, 2018. **9**(1): p. 367.

96. Elorza-Vidal, X., Sirisi, S., Gaitán-Peñas, H., Pérez-Rius, C., Alonso-Gardón, M., Armand-Ugón, M., Lanciotti, A., Brignone, M.S., Prat, E., Nunes, V., Ambrosini, E., Gasull, X., and Estévez, R., *GlialCAM/MLC1 modulates LRRC8/VRAC currents in an indirect manner: Implications for megalencephalic leukoencephalopathy.* Neurobiology of Disease, 2018. **119**: p. 88-99.

97.   Hyzinski-Garcia, M.C., Rudkouskaya, A., and Mongin, A.A., *LRRC8A protein is indispensable for swelling-activated and ATP-induced release of excitatory amino acids in rat astrocytes.* J Physiol, 2014. **592**(22): p. 4855-62.

98.   Kubota, K., Kim, J.Y., Sawada, A., Tokimasa, S., Fujisaki, H., Matsuda-Hashii, Y., Ozono, K., and Hara, J., *LRRC8 involved in B cell development belongs to a novel family of leucine-rich repeat proteins.* FEBS Lett, 2004. **564**(1-2): p. 147-52.

99.   Deneka, D., Sawicka, M., Lam, A.K.M., Paulino, C., and Dutzler, R., *Structure of a volume-regulated anion channel of the LRRC8 family.* Nature, 2018. **558**(7709): p. 254-259.

100.   Kefauver, J.M., Saotome, K., Dubin, A.E., Pallesen, J., Cottrell, C.A., Cahalan, S.M., Qiu, Z., Hong, G., Crowley, C.S., Whitwam, T., Lee, W.-H., Ward, A.B., and Patapoutian, A., *Structure of the human volume regulated anion channel.* eLife, 2018. **7**: p. e38461.

101.   Kasuya, G., Nakane, T., Yokoyama, T., Jia, Y., Inoue, M., Watanabe, K., Nakamura, R., Nishizawa, T., Kusakizako, T., Tsutsumi, A., Yanagisawa, H., Dohmae, N., Hattori, M., Ichijo, H., Yan, Z., Kikkawa, M., Shirouzu, M., Ishitani, R., and Nureki, O., *Cryo-EM structures of the human volume-regulated anion channel LRRC8.* Nat Struct Mol Biol, 2018. **25**(9): p. 797-804.

102.   Kern, D.M., Oh, S., Hite, R.K., and Brohawn, S.G., *Cryo-EM structures of the DCPIB-inhibited volume-regulated anion channel LRRC8A in lipid nanodiscs.* eLife, 2019. **8**: p. e42636.

103.   Schober, A.L., Wilson, C.S., and Mongin, A.A., *Molecular composition and heterogeneity of the LRRC8-containing swelling-activated osmolyte channels in primary rat astrocytes.* The Journal of Physiology, 2017. **595**(22): p. 6939-6951.

104.   Syeda, R., Qiu, Z., Dubin, Adrienne E., Murthy, Swetha E., Florendo, Maria N., Mason, Daniel E., Mathur, J., Cahalan, Stuart M., Peters, Eric C., Montal, M., and Patapoutian, A., *LRRC8 Proteins Form Volume-Regulated Anion Channels that Sense Ionic Strength.* Cell, 2016. **164**(3): p. 499-511.

105.   König, B., Hao, Y., Schwartz, S., Plested, A.J.R., and Stauber, T., *A FRET sensor of C-terminal movement reveals VRAC activation by plasma membrane DAG signaling rather than ionic strength.* eLife, 2019. **8**: p. e45421.

106.   Kumar, L., Chou, J., Yee, C.S., Borzutzky, A., Vollmann, E.H., von Andrian, U.H., Park, S.Y., Hollander, G., Manis, J.P., Poliani, P.L., and Geha, R.S., *Leucine-rich repeat containing 8A (LRRC8A) is essential for T lymphocyte development and function.* J Exp Med, 2014. **211**(5): p. 929-42.

107.   Platt, C.D., Chou, J., Houlihan, P., Badran, Y.R., Kumar, L., Bainter, W., Poliani, P.L., Perez, C.J., Dent, S.Y.R., Clapham, D.E., Benavides, F., and Geha, R.S., *Leucine-rich repeat containing 8A (LRRC8A)-dependent volume-regulated anion channel activity is dispensable for T-cell development and function.* J Allergy Clin Immunol, 2017. **140**(6): p. 1651-1659.e1.

108.   Stuhlmann, T., Planells-Cases, R., and Jentsch, T.J., *LRRC8/VRAC anion channels enhance beta-cell glucose sensing and insulin secretion.* Nat Commun, 2018. **9**(1): p. 1974.

109.   Lück, J.C., Puchkov, D., Ullrich, F., and Jentsch, T.J., *LRRC8/VRAC anion channels are required for late stages of spermatid development in mice.* Journal of Biological Chemistry, 2018. **293**(30): p. 11796-11808.

110.   Bao, J., Perez, C.J., Kim, J., Zhang, H., Murphy, C.J., Hamidi, T., Jaubert, J., Platt, C.D., Chou, J., Deng, M., Zhou, M.H., Huang, Y., Gaitan-Penas, H., Guenet, J.L.,

Lin, K., Lu, Y., Chen, T., Bedford, M.T., Dent, S.Y., Richburg, J.H., Estevez, R., Pan, H.L., Geha, R.S., Shi, Q., and Benavides, F., *Deficient LRRC8A-dependent volume-regulated anion channel activity is associated with male infertility in mice*. JCI Insight, 2018. **3**(16).

111. Lu, J., Xu, F., and Zhang, J., *Inhibition of angiotensin II-induced cerebrovascular smooth muscle cell proliferation by LRRC8A downregulation through suppressing PI3K/AKT activation*. Human Cell, 2019.

112. Rubino, S., Bach, M.D., Schober, A.L., Lambert, I.H., and Mongin, A.A., *Downregulation of Leucine-Rich Repeat-Containing 8A Limits Proliferation and Increases Sensitivity of Glioblastoma to Temozolomide and Carmustine*. Frontiers in oncology, 2018. **8**: p. 142-142.

113. Sirianant, L., Wanitchakool, P., Ousingsawat, J., Benedetto, R., Zormpa, A., Cabrita, I., Schreiber, R., and Kunzelmann, K., *Non-essential contribution of LRRC8A to volume regulation*. Pflugers Arch, 2016. **468**(5): p. 805-16.

114. Galietta, L.J.V., Haggie, P.M., and Verkman, A.S., *Green fluorescent protein-based halide indicators with improved chloride and iodide affinities*. FEBS Letters, 2001. **499**(3): p. 220-224.

115. Decher, N., Lang, H.J., Nilius, B., Brüggemann, A., Busch, A.E., and Steinmeyer, K., *DCPIB is a novel selective blocker of ICl,swell and prevents swelling-induced shortening of guinea-pig atrial action potential duration*. British Journal of Pharmacology, 2001. **134**(7): p. 1467-1479.

116. Ertongur-Fauth, T., Hochheimer, A., Buescher, J.M., Rapprich, S., and Krohn, M., *A novel TMEM16A splice variant lacking the dimerization domain contributes to calcium-activated chloride secretion in human sweat gland epithelial cells*. Exp Dermatol, 2014. **23**(11): p. 825-31.

117. Hamann, S., Kiilgaard, J.F., Litman, T., Alvarez-Leefmans, F.J., Winther, B.R., and Zeuthen, T., *Measurement of Cell Volume Changes by Fluorescence Self-Quenching*. Journal of Fluorescence, 2002. **12**(2): p. 139-145.

118. Liu, T. and Stauber, T., *The Volume-Regulated Anion Channel LRRC8/VRAC Is Dispensable for Cell Proliferation and Migration*. International Journal of Molecular Sciences, 2019. **20**(11): p. 2663.

119. Szücs, G., Heinke, S., Droogmans, G., and Nilius, B., *Activation of the volume-sensitive Cl− current in vascular endothelial cells requires a permissive intracellular Ca2+ concentration*. Vol. 431. 1996. 467-9.

120. Ponomarchuk, O., Boudreault, F., Orlov, S.N., and Grygorczyk, R., *Calcium is not required for triggering volume restoration in hypotonically challenged A549 epithelial cells*. Pflügers Archiv - European Journal of Physiology, 2016. **468**(11): p. 2075-2085.

121. Altamirano, J., Brodwick, M.S., and Alvarez-Leefmans, F.J., *Regulatory volume decrease and intracellular Ca2+ in murine neuroblastoma cells studied with fluorescent probes*. The Journal of general physiology, 1998. **112**(2): p. 145-160.

122. Miriel, V.A., Mauban, J.R.H., Blaustein, M.P., and Gil Wier, W., *Local and cellular Ca2+ transients in smooth muscle of pressurized rat resistance arteries during myogenic and agonist stimulation*. The Journal of Physiology, 1999. **518**(3): p. 815-824.

123. Berridge, M.V., Herst, P.M., and Tan, A.S., *Tetrazolium dyes as tools in cell biology: New insights into their cellular reduction*, in *Biotechnology Annual Review*. 2005, Elsevier. p. 127-152.

124. Gratzner, H., *Monoclonal antibody to 5-bromo- and 5-iododeoxyuridine: A new reagent for detection of DNA replication*. Science, 1982. **218**(4571): p. 474-475.

125.   Coulombe, P.A., Kopan, R., and Fuchs, E., *Expression of keratin K14 in the epidermis and hair follicle: insights into complex programs of differentiation.* The Journal of cell biology, 1989. **109**(5): p. 2295-2312.

126.   Byrne, C., Tainsky, M., and Fuchs, E., *Programming gene expression in developing epidermis.* Development, 1994. **120**(9): p. 2369-2383.

127.   Kartasova, T., Roop, D.R., and Yuspa, S.H., *Relationship between the expression of differentiation-specific keratins 1 and 10 and cell proliferation in epidermal tumors.* Molecular Carcinogenesis, 1992. **6**(1): p. 18-25.

128.   Lanzafame, M., Botta, E., Teson, M., Fortugno, P., Zambruno, G., Stefanini, M., and Orioli, D., *Reference genes for gene expression analysis in proliferating and differentiating human keratinocytes.* Experimental Dermatology, 2015. **24**(4): p. 314-316.

129.   Deyrieux, A.F. and Wilson, V.G., *In vitro culture conditions to study keratinocyte differentiation using the HaCaT cell line.* Cytotechnology, 2007. **54**(2): p. 77-83.

130.   Lenoir, M.C. and Bernard, B.A., *Architecture of reconstructed epidermis on collagen lattices varies according to the method used: a comparative study.* Skin Pharmacol, 1990. **3**(2): p. 97-106.

131.   Brandner, J.M., *Importance of Tight Junctions in Relation to Skin Barrier Function.*

132.   Benfenati, V., Caprini, M., Nicchia, G.P., Rossi, A., Dovizio, M., Cervetto, C., Nobile, M., and Ferroni, S., *Carbenoxolone inhibits volume-regulated anion conductance in cultured rat cortical astroglia.* Channels, 2009. **3**(5): p. 323-336.

133.   Zhang, X., Gao, S., Tanaka, M., Zhang, Z., Huang, Y., Mitsui, T., Kamiyama, M., Koizumi, S., Fan, J., Takeda, M., and Yao, J., *Carbenoxolone inhibits TRPV4 channel-initiated oxidative urothelial injury and ameliorates cyclophosphamide-induced bladder dysfunction.* Journal of Cellular and Molecular Medicine, 2017. **21**(9): p. 1791-1802.

134.   Deng, W., Mahajan, R., Baumgarten, C.M., and Logothetis, D.E., *The ICl,swell inhibitor DCPIB blocks Kir channels that possess weak affinity for PIP2.* Pflügers Archiv - European Journal of Physiology, 2016. **468**(5): p. 817-824.

135.   Lv, J., Liang, Y., Zhang, S., Lan, Q., Xu, Z., Wu, X., Kang, L., Ren, J., Cao, Y., Wu, T., Lin, K.L., Yung, K.K.L., Cao, X., Pang, J., and Zhou, P., *DCPIB, an Inhibitor of Volume-Regulated Anion Channels, Distinctly Modulates K2P Channels.* ACS Chemical Neuroscience, 2019.

136.   Fujii, T., Takahashi, Y., Takeshima, H., Saitoh, C., Shimizu, T., Takeguchi, N., and Sakai, H., *Inhibition of gastric H+,K+-ATPase by 4-(2-butyl-6,7-dichloro-2-cyclopentylindan-1-on-5-yl)oxybutyric acid (DCPIB), an inhibitor of volume-regulated anion channel.* European Journal of Pharmacology, 2015. **765**: p. 34-41.

137.   Friard, J., Tauc, M., Cougnon, M., Compan, V., Duranton, C., and Rubera, I., *Comparative Effects of Chloride Channel Inhibitors on LRRC8/VRAC-Mediated Chloride Conductance.* Frontiers in Pharmacology, 2017. **8**(328).

138.   Milenkovic, A., Brandl, C., Milenkovic, V.M., Jendryke, T., Sirianant, L., Wanitchakool, P., Zimmermann, S., Reiff, C.M., Horling, F., Schrewe, H., Schreiber, R., Kunzelmann, K., Wetzel, C.H., and Weber, B.H.F., *Bestrophin 1 is indispensable for volume regulation in human retinal pigment epithelium cells.* Proceedings of the National Academy of Sciences of the United States of America, 2015. **112**(20): p. E2630-E2639.

139.   Sabirov, R.Z., Prenen, J., Tomita, T., Droogmans, G., and Nilius, B., *Reduction of ionic strength activates single volume-regulated anion channels (VRAC) in endothelial cells.* Pflügers Archiv, 2000. **439**(3): p. 315-320.

140. Rugolo, M., Mastrocola, T., De Luca, M., Romeo, G., and Galietta, L.J.V., *A volume-sensitive chloride conductance revealed in cultured human keratinocytes by 36Cl– efflux and whole-cell patch clamp recording.* Biochimica et Biophysica Acta (BBA) - Biomembranes, 1992. **1112**(1): p. 39-44.

141. Netti, V., Pizzoni, A., Pérez-Domínguez, M., Ford, P., Pasantes-Morales, H., Ramos-Mandujano, G., and Capurro, C., *Release of taurine and glutamate contributes to cell volume regulation in human retinal Müller cells: differences in modulation by calcium.* Journal of Neurophysiology, 2018. **120**(3): p. 973-984.

142. Liu, Y., Zhang, H., Men, H., Du, Y., Xiao, Z., Zhang, F., Huang, D., Du, X., Gamper, N., and Zhang, H., *Volume-regulated Cl- current: contributions of distinct Cl- channel and localized Ca2+ signals.* American Journal of Physiology-Cell Physiology. **0**(0): p. null.

143. Ong, H.L., Subedi, K.P., Son, G.Y., Liu, X., and Ambudkar, I.S., *Tuning store-operated calcium entry to modulate Ca(2+)-dependent physiological processes.* Biochim Biophys Acta Mol Cell Res, 2019. **1866**(7): p. 1037-1045.

144. Benedetto, R., Sirianant, L., Pankonien, I., Wanitchakool, P., Ousingsawat, J., Cabrita, I., Schreiber, R., Amaral, M., and Kunzelmann, K., *Relationship between TMEM16A/anoctamin 1 and LRRC8A.* Pflügers Archiv - European Journal of Physiology, 2016. **468**(10): p. 1751-1763.

145. D'Alessandro, M., Russell, D., Morley, S.M., Davies, A.M., and Lane, E.B., *Keratin mutations of epidermolysis bullosa simplex alter the kinetics of stress response to osmotic shock.* Journal of Cell Science, 2002. **115**(22): p. 4341-4351.

146. Blase, C., Becker, D., Kappel, S., and Bereiter-Hahn, J., *Microfilament dynamics during HaCaT cell volume regulation.* European Journal of Cell Biology, 2009. **88**(3): p. 131-139.

147. He, D., Luo, X., Wei, W., Xie, M., Wang, W., and Yu, Z., *DCPIB, A Specific Inhibitor of Volume-Regulated Anion Channels (VRACs), Inhibits Astrocyte Proliferation and Cell Cycle Progression Via G1/S Arrest.* Journal of Molecular Neuroscience, 2012. **46**(2): p. 249-257.

148. Tao, R., Lau, C.-P., Tse, H.-F., and Li, G.-R., *Regulation of cell proliferation by intermediate-conductance Ca2+-activated potassium and volume-sensitive chloride channels in mouse mesenchymal stem cells.* American Journal of Physiology-Cell Physiology, 2008. **295**(5): p. C1409-C1416.

149. Shen, M.R., Droogmans, G., Eggermont, J., Voets, T., Ellory, J.C., and Nilius, B., *Differential expression of volume-regulated anion channels during cell cycle progression of human cervical cancer cells.* The Journal of physiology, 2000. **529 Pt 2**(Pt 2): p. 385-394.

150. Wong, R., Chen, W., Zhong, X., Rutka, J.T., Feng, Z.-P., and Sun, H.-S., *Swelling-induced chloride current in glioblastoma proliferation, migration, and invasion.* Journal of Cellular Physiology, 2018. **233**(1): p. 363-370.

151. Chen, L., Becker, T.M., Koch, U., and Stauber, T., *The LRRC8/VRAC anion channel facilitates myogenic differentiation of murine myoblasts by promoting membrane hyperpolarization.* Journal of Biological Chemistry, 2019. **294**(39): p. 14279-14288.

152. Wondergem, R., Gong, W., Monen, S.H., Dooley, S.N., Gonce, J.L., Conner, T.D., Houser, M., Ecay, T.W., and Ferslew, K.E., *Blocking swelling-activated chloride current inhibits mouse liver cell proliferation.* The Journal of Physiology, 2001. **532**(3): p. 661-672.

153. Voets, T., Wei, L., Smet, P.D., Driessche, W.V., Eggermont, J., Droogmans, G., and
      Nilius, B., *Downregulation of volume-activated Cl- currents during muscle
      differentiation.* American Journal of Physiology-Cell Physiology, 1997. **272**(2): p.
      C667-C674.

154. Ryle, C.M., Breitkreutz, D., Stark, H.-J., Fusening, N.E., Leigh, I.M., Stelnert, P.M.,
      and Roop, D., *Density-dependent modulation of synthesis of keratins 1 and 10 in the
      human keratinocyte line HACAT and in ras-transfected tumorigenic clones.*
      Differentiation, 1989. **40**(1): p. 42-54.

155. Micallef, L., Belaubre, F., Pinon, A., Jayat-Vignoles, C., Delage, C., Charveron, M.,
      and Simon, A., *Effects of extracellular calcium on the growth-differentiation switch in
      immortalized keratinocyte HaCaT cells compared with normal human keratinocytes.*
      Experimental Dermatology, 2009. **18**(2): p. 143-151.

156. Capone, A., Visco, V., Belleudi, F., Marchese, C., Cardinali, G., Bellocci, M.,
      Picardo, M., Frati, L., and Torrisi, M.R., *Up-Modulation of the Expression of
      Functional Keratinocyte Growth Factor Receptors Induced by High Cell Density in the
      Human Keratinocyte HaCaT Cell Line.* Cell Growth Differ, 2000. **11**(11): p. 607-614.

157. Banks-Schlegel, S. and Green, H., *Involucrin synthesis and tissue assembly by
      keratinocytes in natural and cultured human epithelia.* The Journal of Cell Biology,
      1981. **90**(3): p. 732-737.

158. Watt, F.M. and Green, H., *Involucrin synthesis is correlated with cell size in human
      epidermal cultures.* The Journal of Cell Biology, 1981. **90**(3): p. 738-742.

159. Sun, T.-T. and Green, H., *Differentiation of the epidermal keratinocyte in cell culture:
      Formation of the cornified envelope.* Cell, 1976. **9**(4): p. 511-521.

160. Watt, F.M., *Influence of Cell Shape and Adhesiveness on Stratification and Terminal
      Differentiation of Human Keratinocytes in Culture.* Journal of Cell Science, 1987.
      1987(Supplement 8): p. 313-326.

161. Guo, M., Pegoraro, A.F., Mao, A., Zhou, E.H., Arany, P.R., Han, Y., Burnette, D.T.,
      Jensen, M.H., Kasza, K.E., Moore, J.R., Mackintosh, F.C., Fredberg, J.J., Mooney,
      D.J., Lippincott-Schwartz, J., and Weitz, D.A., *Cell volume change through water
      efflux impacts cell stiffness and stem cell fate.* Proceedings of the National Academy of
      Sciences, 2017. **114**(41): p. E8618-E8627.

162. Janes, S.M., Ofstad, T.A., Campbell, D.H., Watt, F.M., and Prowse, D.M., *Transient
      activation of FOXN1 in keratinocytes induces a transcriptional programme that promotes
      terminal differentiation: contrasting roles of FOXN1 and Akt.* Journal of Cell Science,
      2004. **117**(18): p. 4157-4168.

163. Kim, D., Langmead, B., and Salzberg, S.L., *HISAT: a fast spliced aligner with low
      memory requirements.* Nat Methods, 2015. **12**(4): p. 357-60.

164. Li, H., Handsaker, B., Wysoker, A., Fennell, T., Ruan, J., Homer, N., Marth, G.,
      Abecasis, G., Durbin, R., and Genome Project Data Processing, S., *The Sequence
      Alignment/Map format and SAMtools.* Bioinformatics, 2009. **25**(16): p. 2078-9.

165. Brinkman, E.K., Chen, T., Amendola, M., and van Steensel, B., *Easy quantitative
      assessment of genome editing by sequence trace decomposition.* Nucleic Acids Research,
      2014. **42**(22): p. e168-e168.

166. De Coster, W., D'Hert, S., Schultz, D.T., Cruts, M., and Van Broeckhoven, C.,
      NanoPack: visualizing and processing long-read sequencing data. Bioinformatics, 2018.
      34(15): p. 2666-2669.

167. Li, H., *Minimap2: pairwise alignment for nucleotide sequences.* Bioinformatics, 2018.
      34(18): p. 3094-3100.

10 Appendix

10.1 Abbreviations

bp	Base pair
BrdU	5-bromo-2'-deoxyuridine
CaCC	Calcium-activated chloride channel
Cas	CRISPR associated protein
CBX	Carbenoxolone; 3β-Hydroxy-11-oxoolean-12-en-30-oic acid 3-hemisuccinate
CDS	Coding sequence
CRISPR	Clustered regularly interspaced short palindromic repeats
DAPI	4',6-Diamidin-2-phenylindol
DCPIB	(4-[(2-Butyl-6,7-dichloro-2-cyclopentyl-2,3-dihydro-1-oxo-1H-inden-5-yl)oxy]butanoic acid)
DIDS	Disodium 4,4'-Diisothiocyanato-2,2'-stilbenedisulfonate
DMEM	Dulbecco's modified eagle's medium
EDTA	Ethylenediaminetetraacetic acid
EGTA	Ethyleneglycolbisaminoethylethertetraacetic acid
FCS	Fetal calf serum
FPKM	Fragments per kilobase of exon per million fragments mapped
fw	Forward
GAPDH	Glyceraldehyde-3-phosphate dehydrogenase
gDNA	Genomic DNA
HEK293	Human embryonic kidney cells 293
HEPES BSS	2-[4-(2-hydroxyethyl)piperazin-1-yl]ethanesulfonic acid buffered balanced salt solution
hsYFP	Halide-sensitive yellow fluorescent protein
Hypo	Hypotonic
IC_{50}	Half-maximal inhibitory concentration
IHC	Immunohistochemistry
Iso	Isotonic
kDa	kilo Dalton
KO	Knock-out
KSC	Keratinocyte stem cell
LRRC	Leucine-rich repeat containing protein
MOI	Multiplicity of infection
NFA	Niflumic acid; 2-[3-(Trifluoromethyl)anilino]nicotinic acid
NHEK	Normal human epidermal keratinocyte
ns	Not significant
OFP	Orange fluorescent protein
PBS	Phosphate buffered saline
PCR	Polymerase chain reaction
PM	Post-mitotic cell
rev	Reverse
RPLP0	Ribosomal protein lateral stalk subunit P0
rpm	Rounds per minute

RT	Room temperature
RVD	Regulatory volume decrease
RVI	Regulatory volume increase
SD	Standard deviation
sgRNA	Single-guide RNA
SOCE	Store-operated Ca^{2+} entry
TA	Transient amplifying cell
TMEM	Transmembrane protein
TrypLE	Trysin-like enzyme
UV	Ultraviolet
VRAC	Volume-regulated anion channel
WB	Western blot
WST	Water soluble tetrazolium
WT	Wild-type
YWHAZ	Tyrosine 3-monooxygenase/tryptophan 5-monooxygenase activation protein zeta

10.2 Ehrenwörtliche Erklärung

Ich erkläre hiermit ehrenwörtlich, dass ich die vorliegende Arbeit entsprechend den Regeln guter wissenschaftlicher Praxis selbstständig und ohne unzulässige Hilfe Dritter angefertigt habe.

Sämtliche aus fremden Quellen direkt oder indirekt übernommenen Gedanken sowie sämtliche von Anderen direkt oder indirekt übernommenen Daten, Techniken und Materialien sind als solche kenntlich gemacht. Die Arbeit wurde bisher bei keiner anderen Hochschule zu Prüfungszwecken eingereicht.

10.3 *Curriculum vitae*

Personal information

Name   Janina Trothe
Date of birth   August 10, 1989
Place of birth   Darmstadt

Academic education

Feb 2016 – Jan 2020   PhD candidate supervised by Dr. T. Ertongur-Fauth, BRAIN AG
                        Zwingenberg and Prof. Dr. H. U. Göringer, Institute of Molecular
                        Genetics at Technische Universität Darmstadt

                        PhD Thesis: Importance of volume-regulated anion channel
                        subunit LRRC8A for hypotonic stress response and differentiation
                        of human keratinocytes

Sept 2013 – Sept 2015   Master of Science program "Biomedical Science and Technology"
                        University of Applied Sciences Mannheim

                        Master Thesis: Characterization of volume-regulated anion
                        channels in human skin cells
                        Supervised by Dr. T. Ertongur-Fauth, BRAIN AG Zwingenberg

Mar 2010 – July 2013   Bachelor of Science program "Biological Chemistry"
                        University of Applied Sciences Mannheim

                        Bachelor Thesis: Amyloid and tau pathology in the olfactory
                        epithelium of Alzheimer's disease patients
                        supervised by Prof. Dr. A. Schneider, German Centre for
                        Neurodegenerative Diseases DZNE Göttingen

Aug 1999 – May 2008   Secondary school
                        Lichtenbergschule Darmstadt
                        University-entrance qualification

Publications

Trothe, J., Ritzmann, D., Lang, V., Scholz, P., Pul, U., Kaufmann, R., Buerger, C., and
Ertongur-Fauth, T., *Hypotonic stress response of human keratinocytes involves LRRC8A as
component of volume-regulated anion channels.* Exp Dermatol, 2018. **27**(12): p. 1352-1360.

Pellkofer, H., Ihler, F., Weiss, B.G., Trothe, J., Kadavath, H., Chongtham, M., Kunadt, M.,
Riedel, D., Lornsen, F., Wilken, P., Bartels, C., Hirschel, S., Russo, S.G., Stransky, E.,
Trojan, L., Schmidt, B., Mandelkow, E., Zweckstetter, M., Canis, M., and Schneider, A.,
*Evaluation of the methoxy-X04 derivative BSC4090 for diagnosis of prodromal and early
Alzheimer's disease from bioptic olfactory mucosa.* Eur Arch Psychiatry Clin Neurosci, 2018

10.4 Danksagung

Viele Menschen waren direkt oder indirekt an dieser Arbeit beteiligt, denen ich hier danken möchte.

Vielen Dank, Dr. Torsten Ertongur-Fauth, für die Vergabe des Themas, die Betreuung dieser Arbeit und das Vertreten dieses „Projektes" gegenüber der Brain.

Vielen Dank, Prof. Dr. Uli Göringer, für deine Begleitung mit nützlichen Denkanstößen und kritischem Hinterfragen und die Begutachtung dieser Arbeit.

Vielen Dank, Prof. Dr. Gerhard Thiel, für die Übernahme des Zweitgutachtens, hilfreiche Anregungen und das Voranbringen dieses Themas in der eigenen Arbeitsgruppe.

Mein Dank gilt vielen Kollegen der Brain AG, die mich unterstützt haben. Danke, Dirk R., für deine Hilfe bei den Messungen der Inhibitoren, jegliche Aushilfe mit Medium, Plasmiden und mikrobiologischem Wissen und deine unkomplizierte Art. Danke, Isabella, für deine geleistete Arbeit, die die Proliferation der Keratinozyten untersucht hat. Danke an Paul und Jan, für die bioinformatische Auswertung der NGS und Nanopore Daten und euer unermüdliches Bemühen, mir dies zu erklären. Danke, Ümit, für deine Hilfe und Unterstützung bezüglich aller Fragen zu CRISPR. Danke an das Team der Zellkultur, für eine entspannte und kollegiale Zusammenarbeit und an jeden Kollegen, der mit Interesse und kritischen Rückfragen zur Entwicklung dieser Arbeit beigetragen hat. Danke der Brain AG für die Ermöglichung meine Doktorarbeit durch Finanzierung mit eigenen Mitteln und im Rahmen der NatLifE2020 (FKZ031B0089A). Mein größter Dank geht an Torsten, für deinen Mut und deine Hingabe bei der Betreuung meiner Doktorarbeit, unzählige konstruktive und bereichernde Diskussionen, deine unermüdliche Begeisterung für Wissenschaft und die Begleitung auf dem Weg zu der Wissenschaftlerin, die ich heute bin.

Ein großes Dankeschön geht an meine Kollegen der Uniklinik Frankfurt, für das Beibringen eurer Methoden und das Teilen eurer Antikörperliste ;) Danke, Claudia, für eine erfolgreiche Zusammenarbeit an Publikation und Patent, die Möglichkeit mit und bei euch zu arbeiten und tolle gemeinsame Konferenzbesuche. Danke, Victoria, für deine immunhistochemischen Färbungen, Hilfe bei Methoden und deine direkte Art.

Über diese Doktorarbeit hinaus geht ein Dank an meine langjährigen Freunde Regina, Maxi, Thomas, Katharina und Sabine, für Ablenkung, wenn mir die Decke auf den Kopf gefallen ist, viele offene Ohren und ehrliche Meinungen und dafür, dass ich mich auf euch verlassen kann. Ein großes Dankeschön geht an Susi und Andi, dafür, dass ihr mir einen Ort gegeben habt, an dem der Großteil dieser Arbeit zu Papier gebracht wurde, die Möglichkeit der perfekten Mittagspause und eure ganz eigene Form der Begleitung.

Der größte Dank geht an meine Familie. Herzlichen Dank, Mama und Papa, für eure lebenslange Begleitung, dafür, dass ihr immer an mich glaubt, mir alles zutraut und mich auf allen Wegen unterstützt. Danke für so viel mehr, was ich nicht beschreiben kann. Ohne euch wäre ich diesen Weg nie gegangen.